FOOD ADDITIVES

Edited by **Desiree Nedra Karunaratne**
and **Geethi Pamunuwa**

Food Additives
http://dx.doi.org/10.5772/65204
Edited by Desiree Nedra Karunaratne and Geethi Pamunuwa

Contributors
Muhammad Sajid Arshad, Syeda Ayesha Batool, Dorota Kregiel, Hubert Antolak, Vania Paschoalin, Caroline Joy Steel, Luis Carlos Gioia Jr., José Ricardo Crepaldi Ganancio, Galina Novik, Olga Meerovskaya, Victoria Savich, Desiree Nedra Karunaratne

Published by InTech
Janeza Trdine 9, 51000 Rijeka, Croatia

Publishing Process Manager Dajana Pemac
Technical Editor SPi Global
Cover InTech Design team

First published August, 2017
Printed in Croatia
Legal deposit, Croatia: National and University Library in Zagreb

Additional hard copies can be obtained from orders@intechopen.com

Food Additives , Edited by Desiree Nedra Karunaratne and Geethi Pamunuwa
p. cm.
Print ISBN 978-953-51-3489-3
Online ISBN 978-953-51-3490-9

For EU product safety concerns:
IN TECH d.o.o., Prolaz Marije Krucifikse Kozulić 3, 51000 Rijeka, Croatia,
info@intechopen.com or visit our website at intechopen.com.

Meet the editors

Nedra Karunaratne is a professor of Chemistry at the University of Peradeniya and has obtained her BSc degree (Chem. Hons.) from the University of Colombo, Sri Lanka, and her PhD degree from the University of British Columbia, Canada. Her research covers several areas such as nanotechnology applications of polymer nanoparticles and liposomes for delivery of active ingredients in pharmaceutics, cosmetics and nutraceuticals in food. In addition, she has experience in structural determinations of bacterial polysaccharides and has searched for natural products with antioxidant, antibacterial and anticancer activity.
She has over 50 publications in peer-reviewed journals, 7 patents, 12 book chapters and 97 presentations at conferences. She has edited several books for InTechOpen.

Geethi Pamunuwa obtained her BSc degree (Chem. Hons.) from the University of Peradeniya in 2006; MA degree (Chemistry) from Wayne State University, USA, in 2010; and PhD degree (Chemical Sciences) from the University of Peradeniya, Sri Lanka, in 2015. Her research interests include transdermal delivery of bioactive agents, liposomes, polymer nanoparticles, food formulation and preservation and biopesticides. She currently works as a senior lecturer in food science and technology at the Department of Horticulture and Landscape Gardening, Wayamba University of Sri Lanka. She has published five articles in peer-reviewed journals, one book chapter, and has made seven presentations at various conferences.

Contents

Preface

Food additives have currently become an integral part of the food sector all over the world. In fact, continuously advancing food processing and preservation methods add much emphasis on the need for food additives with diverse attributes. This book, consisting of three sections, contributes significantly to the academic arena and food industry on food additives currently used in the world. The three sections of this monograph are 'Properties of food additives', 'Additives from plant origin' and 'Additives from microbial origin'.

The first section 'Properties of food additives' comprises the introductory chapter 'Introduction to Food Additives' authored by the two editors. This chapter discusses the different types of food additives according to their E-numbers (international numbering system). Intriguingly, this piece of writing indicates the various means of enhancing the properties of the different types of food additives. While pointing out the newest trend of the inclination towards natural food additives by the consumers mainly due to such natural food additives exhibiting much less side effects to the human beings, this chapter highlights the effort taken by the governing bodies worldwide to regularise the use of food additives mainly to mitigate or eradicate the adverse effects of almost all food additives.

The second section is 'Additives from plant origin,' which comprises three chapters. This section begins with the chapter which elaborates the potency of beetroot formulations as a source of, mainly, nitrate ions. The authors give a comprehensive account of the studies carried out so far on the health effects of different beetroot formulations that may be utilised as food additives. Also, this chapter provides information on other important bioactive agents, such as betalains, phenolic compounds and saponins, not restricting its scope only to nitrate ions and their benefits. The authors of the second chapter focus on plant secondary metabolites and bioactive agents as an entrée to the chapter. After bringing in the concept of 'functional foods', they highlight the potential utility of fruit juices and herbal extracts, including those having negative sensory attributes, in the food industry. This chapter expresses the significance of numerous commonly used and highly valued herbs and spices such as peppermint, basil, rosemary, thyme, nettle, elder flowers, cinnamon bark, cloves and liquorice. The discussion of each herb or spice expands incorporating rigorous accounts of not only their antimicrobial activities but also their uses, bioactive compounds and health benefits. The antimicrobial activity is emphasised with respect to the affected bacterial and fungal species. The chapter concludes with uses of fruit juices from different berries as food preservatives and food colours. The bioactive compounds, health benefits and antimicrobial effects of these berries are elaborated. The antioxidant and flavouring properties of plant antimicrobials that benefit in preserving meat and meat-based products are described in the third chapter. The authors focused on the effect of these antimicrobial compounds on bacterial species such as *Salmonella* spp.,*Campylobacter* spp., *Listeria monocytogenes, Escherichia coli* and *Staphylococcus aureus,* found in meat and meat-based products. Impressively, this chapter gives light to the mechanism of action and the factors affecting the activity of plant-derived antimicrobials in addition to relating the active compounds to their target micro-organisms. Also, combined effects of some antimicrobials are emphasised. Herbs and spices and fruits and vegetables are discussed as the sources of plant antimicrobials. Interestingly, the authors then explain the use of antimicrobial edible coating in meat products. The use of multiple methods of preservation of meat is also briefly discussed.

Clean labelling is a concept that the food manufacturers and sellers desire to adhere as a consequence of the consumer preference for natural ingredients. In fact, the general public is much knowledgeable about the health effects of both artificial and natural food additives nowadays and, thus, shows much partiality to natural food additives. Discussing systematically about different beetroot formulations, food preservatives from plants and natural antimicrobials, the three chapters of the second section of this book reveal much valuable information, especially to the different parties in the food industry and academic world in quest of knowledge on food additives from plant origin.

The third and the final section of this book is 'Additives from microbial origin'. This section begins with the chapter in which the writers give a brief yet informative account of lactic acid bacteria including the methods of identification, physiology and metabolism at the very beginning of the chapter. Applications of lactic acid bacteria (LAB) in the food industry are discussed stressing that this additive has gained 'Generally Recognized as Safe' by the Food and Agriculture Organization of the United Nations (FAO). The utility of LAB in fermented food products including dairy and beverage is then explained. An account of the use of LAB as probiotics, disease-suppressing agents and targeted delivery facilitators follows. Another remarkable property of LAB is its antifungal properties that are stressed in this chapter. This interesting chapter then elaborates the use of LAB in waste degradation to produce useful products and recovery of useful components from different food processes. The last chapter of the last section of this book commences with an informative overview of the reactions occurring and the need for food additives in bread-making. Although categorised under this section, this chapter discusses not only of microbially derived additives but also of other food additives used in bread-making. The authors first talk about additives in bread-making, and they disclose, in general, their chemistry, the effects on the dough or bread-making and the dosages. The writers write about 'fermentates' as microbially derived additives. Next, the authors discuss enzymes used in bread-making including numerous enzymes from microbes followed by a systematic discussion of the chemistry of the enzymes. In general, the third section is devoted to natural preservatives and, although not to the same extent as the second section, caters to the need of 'clean labelling'.

After an attractive introduction to the food additives, this book elegantly discusses significant areas of the topic under two sections—'Additives from plant origin' and 'Additives from microbial origin'. This book undoubtedly is a much useful resource to the food industry and academic arena.

Prof. Desiree Nedra Karunaratne
Department of Chemistry
University of Peradeniya
Peradeniya, Sri Lanka

Dr. Geethi Pamunuwa
Department of Horticulture and Landscape Gardening
Faculty of Agriculture and Plantation Management
Wayamba University of Sri Lanka
Makandura, Gonawila

Section 1

Properties of Food Additives

Chapter 1

Introductory Chapter: Introduction to Food Additives

Desiree Nedra Karunaratne and
Geethi Kaushalya Pamunuwa

Additional information is available at the end of the chapter

http://dx.doi.org/10.5772/intechopen.70329

1. Preamble

Food additives are utilized in the preparation and processing of almost all types of food in order to give favorable attributes to the food we eat. Very simply, it is a substance which is added to food to enhance its flavor, appearance, or other favorable quality. In fact, the food protection committee of the US national research council defined food additives as "A substance or a mixture of substances other than a basic food stuff that is present in a food as a result of an aspect of production, processing, storage, or packaging" [1]. According to US FDA (Food and Drug Administration), a food additive is "any substance, the intended use of which results or may reasonably be expected to result–directly or indirectly–in its becoming a component or otherwise affecting the characteristics of any food" [2]. Although the term 'food additives' has been used frequently at present, its utilization has been practiced since ancient times; and probably dating back to much earlier than the hunter-gatherer era. Even though food additives confer much benefit to all sectors, such as the manufacturers, retailers, and customers, utilization of food additives must be carried out extremely cautiously.

Additives, for the most part, are synthetic chemicals. Present day consumers are turning to natural ingredients and bio-based additives due to adverse effects caused by some chemicals. Therefore, plant-derived substances are gaining a foot hold as preservatives, colorants, flavors, and even as antibacterial agents [3, 4].

1.1. E numbers (international numbering system) of food additives

Almost all safe-to-use food additives are given 'E numbers' by the European Food Safety Authority. In order to get to this status, the food additive must pass all the safety checks. Following are the general categories of food additives and their E numbers. However, when one food additive has more than one function, it is given only one E number. Chemical compounds and other species are constantly added to the list of safe-to-use food additives as the food additives pass the safety checks. An up to date list of food additives and their E numbers could be obtained from official UK food standards agency web site https://www.food.gov.uk/science/additives/enumberlist#toc-1. The general list of E numbers of food additives is given in **Table 1** [5].

Block of numbers	Food additives
E100-E199	Colors
E200-E299	Preservatives
E300-E399	Antioxidants and acidity regulators
E400-E499	Thickeners, stabilizers and emulsifiers
E500-E599	Anticaking agents
E600-E699	Flavor enhancers
E700-E799	Antibiotics
E900-E999	Glazing agents and sweeteners
E1000-E1599	Additional chemicals

Table 1. E numbers of food additives.

2. Colors

According to the US FDA, "A color additive is any dye, pigment, or substance, which when added or applied to a food, drug or cosmetic, or to the human body, is capable (alone or through reactions with other substances) of imparting color" [2]. Food colors are used as food additives mainly to yield better sensory effects, specifically appearance contentment. The reasons for adding colors to food are manifold. First, color may be lost due to the processing and storage conditions of food, and thus food colors are added to compensate such loss of color. Second, food items with natural colors may show a variation of color, and thus food colors are added to correct such variations in color. Third, food colors may be added to further improve the natural color of the food. Fourth, food colors are added to give color to food items with no color [2].

There are two types of food colors, certified colors and colors exempt from certification. The certified colors are synthetic compounds. They are usually more effective than natural compounds and they do not introduce off-flavors to the foods. Colors derived from natural sources are exempt from certification. These compounds are more expensive than synthetic compounds. Yet, the colors exempt from certification may give off-flavors to the foods [2].

Health effects of food colorants are a major concern among the consumers and regulatory bodies; and thus, carrying out toxicity studies determining health effects are considered very significant today. A recent study revealed that Allura Red AC lacks genotoxicity after the European Food Safety Authority showed its concern on this matter [6]. In addition to toxicity studies, remedies for the adverse effects of food colorants are being evaluated. For example, Rafati et al. demonstrated that the negative effects caused by tartrazine in mice could be mitigated by the simultaneous administration of vitamin E [7].

Although food colors are added to enhance organoleptic appeal of the foods, naturally occurring food colors such as curcumin and riboflavin possess other beneficial health effects. In fact, curcumin exhibits numerous bioactivities such as antioxidant, antimicrobial, and anticancer [8, 9]. Riboflavin, also, acts as an antioxidant, and it is linked to several health benefits [10]. Numerous strategies have been explored to increase the stability of natural colorants due to beneficial health effects or general lack of toxicity of these compounds [11, 12]. As expected, novel sources of natural colorants are being explored due to the positive attributes of natural colorants [13]. In addition, encapsulation techniques and other innovative methods are being explored in order to improve numerous properties of food colorants as opposed to directly add food colorants in food [14].

The list of colors usually used in food manufacturing is stated below [5].

List of colors: Curcumin, Riboflavin, Riboflavin-5′-phosphate, Tartrazine, Quinoline yellow, Sunset Yellow FCF, Orange Yellow S, Cochineal, Carminic acid, Carmines, Azorubine, Carmoisine, Amaranth, Ponceau 4R, Cochineal Red A, Erythrosine, Allura Red AC, Patent Blue V, Indigotine, Indigo Carmine, Brilliant Blue FCF, Chlorophylls and chlorophyllins, Copper complexes of chlorophyll and chlorophyllins, Green S, Plain caramel, Caustic sulphite caramel, Ammonia caramel, Sulphite ammonia caramel, Brilliant Black BN, Black PN, Vegetable carbon, Brown FK, Brown HT, Carotenes, Annatto, Bixin, Norbixin, Paprika extract, Capsanthin, Capsorubin, Lycopene, Beta-apo-8′-carotenal (C30), Ethyl ester of beta-apo-8′-carotenoic acid (C30), Lutein, Canthaxanthin, Beetroot Red, Betanin, Anthocyanins, Litholrubine BK.

3. Preservatives

Food preservatives have become an indispensible part of the food industry today. In simple terms, a food preservative is any substance that hinders food deterioration caused by microbes, enzymes, or any other chemical reaction. Millions of people suffer from hunger as a result of lack of enough food [15] and thus, the advantages of using food preservatives in food processing are plenteous. Food preservatives along with other food additives are under strict control by numerous governing bodies. A short account of the governing system is given under Section 11.

Most artificial food preservatives impart negative health effects at high doses. For instance, *in vitro* studies have revealed that sodium benzoate and potassium benzoate exhibit genotoxic effects [16]. However, this issue can be dealt with by adhering to the acceptable daily intake (ADI) values of food additives (please refer Section 11). Interestingly, despite showing adverse effects at toxic levels, some artificial food preservatives show favorable health effects at nontoxic levels [17].

Natural preservatives are an appealing alternative to artificial preservatives, especially with respect to health effects. A novel trend is to explore and utilize essential oils such as clove essential oil and eugenol extracted from cloves, limonene extracted from citrus fruits, and essential oil extracted from cinnamon as food preservatives of numerous food items including fresh cut produce, juices, and fish [3, 18–20]. As expected, encapsulated natural food preservatives including thyme essential oil and curcumin have shown favorable properties such as sustained release and enhanced antioxidant and antimicrobial properties [21, 22]. In addition to natural products, fermented milk products have shown promise as food preservatives [23]. The reasons for utilizing natural products and nonsynthetic products as food preservatives include imparting health benefits to the consumers and gaining "clean label" advantage.

Numerous approaches are being taken to find novel food preservatives with ameliorated properties. For instance, peptides have been used successfully as potential food preservatives [24]. Once a peptide food preservative is identified, mass production may be carried out using biotechnology. Combinations of food preservatives have also been studied to discover the combined effect and the possibility of substituting synthetic food preservatives by such combinations. For example, *Cuminum cyminum* L. essential oil and nisin have shown their ability to function as a hurdle against microbes [25].

Some food preservatives used in food manufacturing are listed below [5].

List of preservatives: Sorbic acid, Potassium sorbate, Calcium sorbate, Benzoic acid, Sodium benzoate, Potassium benzoate, Calcium benzoate, Ethyl p-hydroxybenzoate, Sodium ethyl p-hydroxybenzoate, Sodium methyl p-hydroxybenzoate, Sulfur dioxide, Sodium sulphite, Sodium hydrogen sulphite, Sodium metabisulphite, Potassium metabisulphite, Calcium sulphite, Calcium hydrogen sulphite, Potassium hydrogen sulphite, Biphenyl; diphenyl, Nisin, Natamycin, Hexamethylene tetramine, Dimethyl dicarbonate, Potassium nitrite, Sodium nitrite, Sodium nitrate, Potassium nitrate, Propionic acid, Sodium propionate, Calcium propionate, Potassium propionate, Boric acid, Sodium tetraborate; borax.

4. Antioxidants and acidity regulators

Antioxidants play a pivotal role in the food industry, combating oxidative stress on oxygen-sensitive species. The antioxidants used in the food industry are either hydrophilic, lipophilic, or amphiphilic, protecting various types of ingredients. Certain antioxidants function also as acidity regulators. Examples include ascorbic acid and citric acid. Acidity regulators are also an essential group of food additives as lowering the pH of the food usually assists to retard microbial attack.

4.1. Antioxidants

Although antioxidants are deemed to confer numerous health benefits to the humans, synthetic antioxidants such as butylated hydroxyanisole (BHA) and butylated hydroxytoluene (BHT) have shown negative health effects [26]. On the contrary, some reports have shown chemoprevention properties of those synthetic carcinogenic antioxidants [27]. Again, the issue of toxicity is dealt with by adhering strictly into the ADI published by the governing bodies worldwide including the US FDA. Although the results of synthetic antioxidants are inconsistent, numerous natural antioxidants have the ability to function as nontoxic anticarcinogenic compounds. Examples include ferulic acid, caffeic acid, curcumin, vitamin E, polyphenolic catechins, and carnosol [28].

As with other food additives, the trend is to utilize and seek for natural food antioxidants. Both pure antioxidants and plant extracts are used and explored these days. Moreover, encapsulation of pure antioxidants and plant extracts showing antioxidant properties is carried out to obtain improved attributes such as improved stability and sustained release of those bioactive compounds [29, 30]. The liposomal encapsulation of the *Schumacheria castaneifolia* methanol extract with antioxidant properties, which may be suitable for applications in the food sector, with high encapsulation efficiencies is an excellent example of encapsulating plant extracts [29].

A list of antioxidants used in food manufacturing is stated below [5].

List of antioxidants: Ascorbic acid, Sodium ascorbate, Calcium ascorbate, Fatty acid esters of ascorbic acid, Tocopherols, Alpha-tocopherol, Gamma-tocopherol, Delta-tocopherol, Propyl gallate, Octyl gallate, Dodecyl gallate, Erythorbic acid, Sodium erythorbate, Tertiary-butyl hydroquinone (TBHQ), Butylated hydroxyanisole (BHA), Butylated hydroxytoluene (BHT), Extracts of rosemary, 4-Hexylresorcinol.

4.2. Acidity regulators

Acidity regulators such as citric acid, tartaric acid, and phosphoric acid are numbered together with antioxidants in the E numbering system [5, 31]. This approach is very logical as certain acidity regulators, such as citric acid, exhibit antioxidant properties. In fact, citric acid has imparted favorable effects on food, functioning as an acidity regulator and antioxidant simultaneously [32]. What's more, food acidity regulators have shown advantageous combined effects with other food additives on food. Antibrowning effect of citric acid together with ascorbic acid and nitrogen on banana smoothies is an example [33].

The list of acidity regulators commonly used in food manufacturing is stated below [5].

List of acidity regulators: Sodium lactate, Potassium lactate, Calcium lactate, Citric acid, Sodium citrates, Potassium citrates, Calcium citrates, Tartaric acid (L-(+)), Sodium tartrates, Potassium tartrates, Sodium potassium tartrate, Phosphoric acid, Sodium phosphates, Potassium phosphates, Calcium phosphates, Magnesium phosphates, Sodium malates, Potassium malate, Calcium malates, Metatartaric acid, Calcium tartrate, Adipic acid, Sodium adipate, Potassium adipate, Succinic acid, Triammonium citrate, Calcium disodium ethylene diamine tetra-acetate; calcium disodium EDTA.

5. Thickeners, stabilizers, emulsifiers, and gelling agents

Thickeners, stabilizers, emulsifiers, and gelling agents have become an integral part in the current food manufacturing industry. Thickeners increase the volume, change the viscosity, and increase the processability of the food items. Stabilizers, as the name implies, stabilize the food products; sometimes through the utilization of fillers. Emulsifiers assist in the miscibility of otherwise immiscible substances possible. For instance water-in-oil or oil-in-water emulsions used in the food industry are made utilizing emulsifiers. Gelling agents mainly contribute to the viscosity and sensory properties of the food products. In sum, all thickeners, stabilizers, emulsifiers and gelling agents contribute to the stability and palatability of the food product.

This category of food additives also consist of natural and synthetic compounds. In fact, lecithin that assists in emulsification and stabilization for most food products is mostly extracted from soy bean, and thus it is a natural additive [34]. However, numerous studies are being conducted evaluating the positive effects of synthetic lecithin [35]. Alginate functioning as both a thickener and gelling agent is another natural food additive in this group [36]. Apart from the natural compounds, synthetic emulsifiers such as polysorbates constitute an important component of this group. Although considered food grade, several health concerns have arisen regarding such artificial emulsifiers [37].

A list of thickeners-stabilizers-emulsifiers-gelling agents used in food manufacturing is listed in **Tables 2** and **3** [5].

List of thickeners-stabilizers-emulsifiers-gelling agents: Lecithins, Alginic acid, Sodium alginate, Potassium alginate, Ammonium alginate, Calcium alginate, Propane-1-2-diol alginate, Agar, Carrageenan, Processed eucheuma seaweed, Locust bean gum; carob gum, Guar gum, Tragacanth, Acacia gum; gum Arabic, Xanthan gum, Karaya gum, Tara gum, Gellan gum, Konjac, Soybean hemicellulose, Cassia gum, Polyoxyethylene sorbitan monolaurate; Polysorbate 20, Polyoxyethylene sorbitan mono-oleate; Polysorbate 80, Polyoxyethylene sorbitan monopalmitate; Polysorbate 40, Polyoxyethylene sorbitan monostearate; Polysorbate 60, Polyoxyethylene sorbitan tristearate; Polysorbate 65, Pectins, Ammonium phosphatides, Sucrose acetate isobutyrate, Glycerol esters of wood rosins, Cellulose, Methyl cellulose, Ethyl cellulose, Hydroxypropyl cellulose, Hydroxypropyl methyl cellulose, Ethyl methyl cellulose, Carboxy methyl cellulose, Crosslinked sodium carboxy methyl cellulose.

6. Anticaking agents

As the name implies, the role of anticaking agents is to prevent lumping or caking in food. These agents are added mostly for powders or granulated material. Among the numerous advantages of using anticaking agents include: sustenance of sensory attributes, easiness of packaging, efficient transportation, and simplicity to yield high quality products for consumption. Depending on the food product involved, either water-soluble or organic solvent-soluble anticaking agents are used.

Anticaking agents frequently used in food manufacturing are stated below [5].

List of anticaking agents: Calcium Aluminum Silicate, Calcium Phosphate tribasic, Calcium Silicate, Calcium Stearate, Cellulose, Magnesium Carbonate, Magnesium Oxide, Magnesium Silicate, Magnesium Stearate, Microcrystalline Cellulose, Propylene Glycol, Potassium Ferrocyanide, Trihydrate, Silicon Dioxide, Sodium Aluminum Silicate, Sodium Ferrocyanide, decahydrate.

7. Flavors and flavor enhancers

Flavors and flavor enhancers are of extreme importance in the food industry as it is what makes the food sensational. Flavor is perceived by the taste and smell via chemical senses. Also, the chemical irritants perceived in the mouth and throat, temperature and texture are factors affecting the flavor of a food. Nowadays, both natural and artificial substances are used as food flavors **Table 2**. The basic universally recognized flavors include: sweet, sour, tangy, bitter, umami, hot, that can be perceived through the tongue. On the other hand, the number of sensations that can be perceived through the nose (smell) is limitless. As a result, the food industry is ever growing utilizing different combinations of taste and smell. What's more, there is another group of chemical substances that do not impart any flavor in to the food product but enhance the existing flavor in the food **Table 3**. These flavor enhances are highly valued in the food industry as these substances contribute significantly into cost reduction in food manufacturing. Flavors and flavor enhancers frequently used in food manufacturing are stated [38].

Flavors and flavor enhances also are evaluated for their health effects by numerous scientists worldwide. Further, extraction of numerous novel natural flavors is being carried out around the globe as a result of the higher inclination of the customers to such natural compounds [39]. There has been much criticism on the health effects of glutamate—a much consumed flavor enhancer. However, mixed results have been published and there is no evidence to prove that glutamate possesses negative health effects, according to a recent report [40]. Like almost all other food additives, encapsulation, for instance microencapsulation and emulsification, is used as means of enhancing the properties of food flavors [41, 42].

Chemical	Odor
Diacetyl, acetylpropionyl, acetoin	Buttery
Isoamyl acetate	Banana
Benzaldehyde	Bitter almond, cherry
Cinnamaldehyde	Cinnamon
Ethyl propionate	Fruity
Methyl anthranilate	Grape
Limonene	Orange
Ethyl decadienoate	Pear
Allyl hexanoate	Pineapple
Ethyl maltol	Sugar, cotton candy
Ethylvanillin	Vanilla
Methyl salicylate	Wintergreen

This table was obtained from: Wikipedia [38].

Table 2. Artificial flavoring agents and their flavors.

Acid	Description
Glutamic acid salts	This amino acid's sodium salt, monosodium glutamate (MSG), is one of the most commonly used flavor enhancers in food processing. Mono- and diglutamate salts are also commonly used.
Glycine salts	Simple amino acid salts typically combined with glutamic acid as flavor enhancers.
Guanylic acid salts	Nucleotide salts typically combined with glutamic acid as flavor enhancers.
Inosinic acid salts	Nucleotide salts created from the breakdown of AMP, due to high costs of production, typically combined with glutamic acid as flavor enhancers.
5′-Ribonucleotide salts	Nucleotide salts typically combined with other amino acids and nucleotide salts as flavor enhancers.

This table was obtained from: Wikipedia [38].

Table 3. Artificial flavor enhancers.

8. Antibiotics

Antibiotics are being used in the food industry today to increase the shelf life of numerous food items, especially perishable food items including milk [43]. Although not directly added during food processing, nonvegetarian food may contain a certain amount of antibiotics since antibiotics are frequently used in animal production. However, any antibiotic used for human therapeutic purposes or for animal feed additive are banned for use in the food industry. Tetracycline is a classic example. Maximum permissible amounts of such antibiotic residues have been declared and much emphasis is given to regular monitoring of antibiotic residues in food [44]. Moreover, the antibiotics used in the food industry show slower activity than those used for therapeutic purposes [45]. Antibiotics frequently used in food manufacturing are stated below.

List of antibiotics: Nisin, Natamycin, Subtilin, Tylosin Phytoncides.

Phytoncides are antibiotics obtained from plants. Examples include: mustard oil, thyme, cinnamaldehyde, eugenol, etc. [46].

Antibiotics permitted as food additives are being experimented heavily, especially to engineer more potent variants [47]. Further, encapsulation has become a common technique to enhance the desirable properties of antibiotics. For instance, coated liposomes encapsulating nisin has shown improved sustained release properties beneficial for applications in the food sector [48].

9. Glazing agents and sweeteners

Glazing agents may be either natural or synthetic. They are used mainly for preservation of food items by forming a thin coat around it [49]. A list of glazing agents frequently used in food industry is stated below [5].

List of glazing agents: Stearic acid, Beeswax, Candelilla wax, Carnauba wax, Shellac, Microcrystalline wax, Crystalline wax, Lanolin, Oxidized polyethylene wax, Esters of colophonium, Paraffin.

The most commonly used sweetener used in the food industry is sucrose as it is readily available. Thus, the performance of other sweeteners is frequently measured against that of sucrose [50]. Glucose is also frequently used in the food industry, especially in the manufacturing of confectionaries [51]. However, substitutes for common sugars, natural or artificial, are in high demand due to the prevalence of diabetes mellitus among a significant proportion of people worldwide. Other requirements for sugar substitutes include weight loss, dental care, and reactive hypoglycemia. In addition, using sugar substitutes is cost effective since the sugar substitutes are many times (sometimes more than 100 or even 1000 times) sweeter than sucrose [52]. A list of sweeteners frequently used in food manufacturing is stated below [5, 53].

List of sweeteners: Sorbitol, Sorbitol syrup, Mannitol, Acesulfame K, Aspartame, Cyclamic acid and its Na and Ca salts, lsomalt, Saccharin and its Na - K and Ca salts, Sucralose, Thaumatin, Neohesperidine DC, Steviol glycoside, Neotame (as a flavor enhancer), Salt of aspartame-acesulfame, Maltitol, Maltitol syrup, Lactitol, Xylitol, Erythritol.

10. Additional chemicals

The European Food Safety Authority has grouped some food additives as 'additional chemicals' as those chemicals cannot be grouped together with other food additives. As indicated in **Table 1**, these chemicals are numbered from E1000 to E1599. Even though these chemicals may function as other food additives, they have different properties and thus treated differently. For instance, invertase having the number E1103 functions as emulsifiers-stabilizers-thickeners-gelling agents but is in a special category.

A list of other chemicals frequently used in food manufacturing is stated below [5].

List of other chemicals: Polydextrose, Polyvinylpyrrolidone, Polyvinylpolypyrrolidone, Polyvinyl alcohol, Pullulan, Basic methacrylate copolymer, Oxidized starch, Monostarch phosphate, Distarch phosphate, Phosphated distarch phosphate, Acetylated distarch phosphate, Acetylated starch, Acetylated distarch adipate, Hydroxyl propyl starch, Hydroxy propyl distarch phosphate, Starch sodium octenyl succinate, Acetylated oxidized starch, Starch aluminum Octenyl succinate, Triethyl citrate, Glyceryl triacetate; triacetin, Propan-1-2-diol; propylene glycol, Polyethylene glycol.

11. Regulation

Food additives are under strict control of numerous governing bodies. In the European Union, the governing bodies are the European Food Safety Authority (EFSA) and the European Commission, Parliament and Council. These bodies are accountable for the safety assessment, which includes toxicological studies and dietary exposure assessment, authorization which includes maintaining and publishing data bases of food additives permitted to be used in the EU, and control which is involved in legislation and labeling of food additives. The U.S. Food and drug administration (US FDA) is the main governing body of food additives in USA, and almost all other countries have their own governing bodies of food safety.

Food and Agriculture Organisation (FAO) and the World Health Organisation (WHO) work together in the international arena via a Joint Expert Committee on Food Additives (JECFA) [54–57].

Joint Expert Committee on Food Additives from 1961 has taken initiative of matters regarding the acceptable daily intake (ADI) level. "ADI is a measure of the amount of a specific substance (originally applied for a food additive, later also for a residue of a veterinary drug or pesticide) in food or drinking water that can be ingested (orally) on a daily basis over a lifetime without an appreciable health risk". "ADIs are expressed usually in milligrams (of the substance) per kilogram of body weight per day." [58]

All of these food additives are used to fine tune the food items to yield a superb food product having sensational attributes. In addition, the preservative effect that the food additives impart is of utmost importance. Further, food safety governing bodies worldwide have set maximum levels to be used in the food industry for all approved food additives. Thus, health risk is at a low level. However, it is advisable to change ones diet time to time so that the subject is not exposed to the same food additives for lengthy periods of time. This practice also may not be essential if the customer pays attention to the recommended daily intake of the ingredients.

12. Take home message

"Innovation is change that unlocks new value" according to Jamie Notter [59]. Adding food additives to enhance the attributes of food is an ancient concept of value addition of food practiced from as early as the hunter-gatherer era. In the modern era, the demand is greater than the supply, and innovation is the change that satisfies the demand by all the sectors, such as the manufacturers, retailers, and customers. While using food additives to enhance the attributes of food, it is of prime importance that guidelines by the relevant food safety authorities are followed since synthetic and natural compounds with various health effects are widely used as additives. Deviations from food safety regulations may result in serious negative outcomes. As a result, any party breaching rules and guidelines regarding food additives will have to face serious consequences, including harsh court decisions against them. In sum, considering the health effects and regulations regarding food additives are extremely significant although utilizing food additives have enabled the beings to enjoy a plethora of various food products.

Author details

Desiree Nedra Karunaratne[1]* and Geethi Kaushalya Pamunuwa[2]

*Address all correspondence to: nedrak@pdn.ac.lk

1 Department of Chemistry, University of Peradeniya, Peradeniya, Sri Lanka

2 Department of Horticulture and Landscape Gardening, Faculty of Agriculture and Plantation Management, Wayamba University of Sri Lanka, Gonawila, Sri Lanka

References

[1] Vaclavik VA, Christian EW. Essentials of Food Science. 3rd ed. Heldman DR, editor. New York, USA: Springer; 2008. ISBN: 978-0-387-69939-4

[2] International Food Information Council (IFIC) and U.S. Food and Drug Administration. Overview of Food Ingredients, Additives & Colors. [Internet] 2010. Available from: https://www.fda.gov/food/ingredientspackaginglabeling/foodadditivesingredients/ucm094211.htm#foodadd. [Accessed: May 28, 2017]

[3] Souza ELD, Almeida ETDC, Guedes JPDS. The potential of the incorporation of essential oils and their individual constituents to improve microbial safety in juices: A review. Comprehensive Reviews in Food Science and Food Safety. 2016;**15**:753-772. DOI: 10.1111/1541-4337.12208

[4] Gómez HC, Serpa A, Velásquez-Cock J, Gañán P, Castro C, Vélez L, Zuluaga, R. Vegetable nanocellulose in food science: A review. Food Hydrocolloids. 2016;**57**:178-186. DOI: 10.1016/j.foodhyd.2016.01.023

[5] Food Additives and Ingredients Association. E numbers. [Internet] 2017. Available from: http://www.faia.org.uk/e-numbers/ [Accessed: April 28, 2017]

[6] Bastaki M, Farrell T, Bhusari S, Pant K, Kulkarni R. Lack of genotoxicity *in vivo* for food color additive Allura Red AC. Food and Chemical Toxicology. 2017;**105**:308-314. DOI: 10.1016/j.fct.2017.04.037

[7] Rafati A, Nourzei N, Karbalay-Doust S, Noorafshan A. Using vitamin E to prevent the impairment in behavioral test, cell loss and dendrite changes in medial prefrontal cortex induced by tartrazine in rats. Acta Histochemica. 2017;**119**:172-180. DOI: 10.1016/j.acthis.2017.01.004

[8] Maheshwari RK, Singh AK, Gaddipati J, Srimal RC. Multiple biological activities of curcumin: A short review. Life Sciences. 2006;**78**(18):2081-2087. DOI: 10.1016/j.lfs.2005.12.007

[9] Wang H, Hao L, Wang P, Chen M, Jiang S, Jiang S. Release kinetics and antibacterial activity of curcumin loaded zein fibers. Food Hydrocolloids. 2017;**63**:437-446. DOI: 10.1016/j.foodhyd.2016.09.028

[10] Mikkelsen K, Stojanovska L, Prakash M, Apostolopoulos V. The effects of vitamin B on the immune/cytokine network and their involvement in depression. Maturitas. 2017;**96**:58-71. DOI: 10.1016/j.maturitas.2016.11.012

[11] Chung C, Rojanasasithara T, Mutilangi W, McClements DJ. Stabilization of natural colors and nutraceuticals: Inhibition of anthocyanin degradation in model beverages using polyphenols. Food Chemistry. 2016;**212**:596-603. DOI: 10.1016/j.foodchem.2016.06.025

[12] Chung C, Rojanasasithara T, Mutilangi W, McClements DJ. Stability improvement of natural food colors: Impact of amino acid and peptide addition on anthocyanin stability in model beverages. Food Chemistry. 2017;**218**:277-284. DOI: 10.1016/j.foodchem.2016.09.087

[13] Vagiri M, Jensen M. Influence of juice processing factors on quality of black chokeberry pomace as a future resource for colour extraction. Food Chemistry. 2017;**217**:409-417. DOI: 10.1016/j.foodchem.2016.08.121

[14] Chittem J, Sajjan GS, Kanumuri MV. Spectrophotometric evaluation of colour stability of nano hybrid composite resin in commonly used food colourants in Asian countries. Journal of Clinical and Diagnostic Research. 2017;**11**:ZC61-ZC65. DOI: 10.7860/JCDR/2017/22919.9193

[15] World hunger and poverty facts and statistics. Hunger Notes, World Hunger Education Service. Fight Hunger with Knowledge. 2016. [Internet] Available from: http://www.worldhunger.org/2015-world-hunger-and-poverty-facts-and-statistics/#hunger-number [Accessed: May 28, 2017]

[16] Zengin N, Yüzbaıoglu D, Unal F, Yilmaz S, Aksoy H. The evaluation of the genotoxicity of two food preservatives: Sodium benzoate and potassium benzoate. Food and Chemical Toxicology. 2011;**49**(4):763-769. DOI: 10.1016/j.fct.2010.11.040

[17] Yadav A, Kumar A, Das M, Tripathi A. Sodium benzoate, a food preservative, affects the functional and activation status of splenocytes at non cytotoxic dose. Food and Chemical Toxicology. 2016;**88**:40-47. DOI: 10.1016/j.fct.2015.12.016

[18] Chen X, Ren L, Li M, Qian J, Fan J, Du B. Effects of clove essential oil and eugenol on quality and browning control of fresh-cut lettuce. Food Chemistry. 2017;**214**:432-439. DOI: 10.1016/j.foodchem.2016.07.101

[19] Giarratana F, Muscolino D, Beninati C, Ziino G, Giuffrida A, Panebianco A. Activity of R(+) limonene on the maximum growth rate of fish spoilage organisms and related effects on shelf-life prolongation of fresh gilthead sea bream fillets. International Journal of Food Microbiology. 2016;**237**:109-113. DOI: 10.1016/j.ijfoodmicro.2016.08.023

[20] Zhang Y, Li D, Lv J, Li Q, Kong C, Luo Y. Effect of cinnamon essential oil on bacterial diversity and shelf-life in vacuum-packaged common carp (*Cyprinus carpio*) during refrigerated storage. International Journal of Food Microbiology. 2017;**249**:1-8. DOI: 10.1016/j.ijfoodmicro.2016.10.008

[21] Asprea M, Leto I, Bergonzi MC, Bilia AR. Thyme essential oil loaded in nanocochleates: Encapsulation efficiency, in vitro release study and antioxidant activity. LWT–Food Science and Technology. 2017;**77**:497-502. DOI: 10.1016/j.lwt.2016.12.006

[22] Gómez-Estacal J, Balaguer MP, López-Carballo G, Gavara R, Hernández-Muñoz P. Improving antioxidant and antimicrobial properties of curcumin by means of encapsulation in gelatin through electrohydrodynamic atomization. Food Hydrocolloids. 2017;**70**:313-320. 10.1016/j.foodhyd.2017.04.019

[23] Gamba RR, Caro CA, Martínez OL, Moretti AF, Giannuzzi L, Antoni GLD, Peláez AL. Antifungal effect of kefir fermented milk and shelf life improvement of corn arepas. International Journal of Food Microbiology. 2016;**235**:85-92. DOI: 10.1016/j.ijfoodmicro.2016.06.038

[24] Li Y-Q, Sun X-X, Feng J-L, Mo H-Z. Antibacterial activities and membrane permeability actions of glycinin basic peptide against *Escherichia coli*. Innovative Food Science and Emerging Technologies. 2015;**31**:170-176. DOI: 10.1016/j.ifset.2015.07.009

[25] Tavakoli HR, Mashak Z, Moradi B, Sodagari HR. Antimicrobial activities of the combined use of *Cuminum cyminum* L. essential oil, nisin and storage temperature against *Salmonella typhimurium* and *Staphylococcus aureus in vitro*. Jundishapur Journal of Microbiology. 2015;**8**(4): e24838. DOI: 10.5812/jjm.8(4)2015.24838

[26] Witschi HP. Enhanced tumour development by butylated hydroxytoluene (BHT) in the liver, lung and gastro-intestinal tract. Food and Chemical Toxicology. 1986;**24**(10-11): 1127-1130. DOI: 10.1016/0278-6915(86)90298-X

[27] Hocman G. Chemoprevention of cancer: Phenolic antioxidants (BHT, BHA). International Journal of Biochemistry. 1988;**20**(7):639-651. DOI: 10.1016/0020-711X(88)90158-9

[28] Huang M-T, Ferraro T. Phenolic compounds in food and cancer prevention. In: Phenolic Compounds in Food and Their Effects on Health II. ACS Symposium Series. 1992;**507**: Chapter 2, pp. 8-34. DOI: 10.1021/bk-1992-0507.ch002

[29] Pamunuwa KMGK, Bandara CJ, Karunaratne V, Karunaratne DN. Optimization of a liposomal delivery system for the highly antioxidant methanol extract of stem-bark of *Schumacheria castaneifolia* Vahl. Chemical and Pharmaceutical Research. 2015;**7**(4):1236-1245. ISSN: 0975-7384

[30] Katuwavila KANP, Perera ADLC, Karunaratne V, Amaratunga GAJ, Karunaratne DN. Improved delivery of caffeic acid through liposomal encapsulation. Journal of Nanomaterials. 2016;**2016**:Article ID 9701870:7. DOI: 10.1155/2016/9701870

[31] Food Standards Agency. Current EU approved additives and their E Numbers. [Internet]. 2015. Available from: https://www.food.gov.uk/science/additives/enumberlist [Accessed: June 09, 2017

[32] Tsouvaltzis P, Brecht JK. Inhibition of enzymatic browning of fresh-cut potato by immersion in citric acid is not solely due to pH reduction of the solution. Journal of Food Processing and Preservation. 2017;**41**(2):e12829. DOI: 10.1111/jfpp.12829

[33] Wang S, Lin T, Man G, Li H, Zhao L, Wu J, Liao X. Effects of anti-browning combinations of ascorbic acid, citric acid, nitrogen and carbon dioxide on the quality of banana smoothies. Food Bioprocess Technology. 2014;**7**(1):161-173. DOI: 10.1007/s11947-013-1107-7.

[34] Lima MSAD, Rocha LA, Molina EF, Caetano BL, Mello LMC, Ciuffi KJ, Calefi PS, Nassar EJ. Thermoanalysis of soybean oil extracted by two methods. Quimica Nova. 2008;**31**(3):527-529. DOI: 10.1590/S0100-40422008000300012

[35] Onaderra M, Monsalve RI, Manchero JCM, Villalba M, Pozo AMD, Gavilanes JCG, Rodriguez R. Food mustard allergen interaction with phospholipid vesicles. European Journal of Biochemistry. 1994;**225**:609-615. DOI: 10.1111/j.1432-1033.1994.00609.x

[36] Rioux LE. Characterization of polysaccharides extracted from brown seaweeds. Carbohydrate Polymers. 2007;**69**(3):530-537. DOI: 10.1016/j.carbpol.2007.01.009

[37] Csáki KF. Synthetic surfactant food additives can cause intestinal barrier dysfunction. Medical Hypotheses. 2011;**76**(5):676-681. DOI: 10.1016/j.mehy.2011.01.030

[38] Wikipedia: The free encyclopedia. Flavor. [Internet] 2017. Available from: https://en.wikipedia.org/wiki/Flavor [Accessed: June 26, 2017]

[39] Alañón ME, Alarcón M, Marchante L, Díaz-Maroto MC, Pérez-Coello MS. Extraction of natural flavorings with antioxidant capacity from cooperage by-products by green extraction procedure with subcritical fluids. Industrial Crops and Products. 2017;**103**:222-232. DOI: 10.1016/j.indcrop.2017.03.050

[40] Jinap S, Hajeb P. Glutamate. Its applications in food and contribution to health. Appetite. 2010;**55**(1):1-10. DOI: 10.1016/j.appet.2010.05.002

[41] Breternitz NR, Bolini HMA, Hubinger MD. Sensory acceptance evaluation of a new food flavoring produced by microencapsulation of a mussel (*Perna perna*) protein hydrolysate. LWT–Food Science and Technology. 2017;**83**:141-149. DOI: 10.1016/j.lwt.2017.05.016

[42] Mao L, Roos YH, Biliaderis CG, Miao S. Food emulsions as delivery systems for flavor compounds: A review. Critical Reviews in Food Science and Nutrition. 2017;**57**(15):3173-3187. DOI: 10.1080/10408398.2015.1098586

[43] Shi C, Zhao X, Meng R, Liu Z, Zhang G, Guo N. Synergistic antimicrobial effects of nisin and p-Anisaldehyde on *Staphylococcus aureus* in pasteurized milk. LWT–Food Science and Technology. 2017;**84**:222-230. DOI: 10.1016/j.lwt.2017.05.056

[44] Rakitsky VN, Sinitskaya TA, ChKhvirkiya EG, Nikolaev NI. Tolerance in the substantiation of the maximum allowable content of antibiotics in food products. Voprosy Pitaniia. 2013;**82**(2):48-52.

[45] Goldstein BP, Wei J, Greenberg K, Novick R. Activity of nisin against *Streptococcus pneumoniae, in vitro,* and in a mouse infection model. Journal of Antimicrobial Chemotherapy. 1998;**42**:277-278. DOI: 10.1093/jac/42.2.277

[46] Burt S. Essential oils: Their antibacterial properties and potential applications in foods—A review. International Journal of Food Microbiology. 2004;**94**:223-252. DOI: 10.1016/j.ijfoodmicro.2004.03.022

[47] Field D, Begley M, O'Connor PM, Daly KM, Hugenholtz F, Cotter PD, Hill C, Ross RP. Bioengineered nisin A derivatives with enhanced activity against both gram positive and gram negative pathogens. PLoS One. 2012;**7**(10):e46884. DOI: 10.1371/journal.pone.0046884

[48] Lopes NA, Pinilla CMB, Brandelli A. Pectin and polygalacturonic acid-coated liposomes as novel delivery system for nisin: Preparation, characterization and release behavior. Food Hydrocolloids. 2017;**70**:1-7. DOI: 10.1016/j.foodhyd.2017.03.016

[49] Pandey RM, Upadhyay SK. Food additive. In: El-Samragy Y, editor. Food Additive. Rijeka: InTech; 2012. pp. 1-30. DOI: 10.5772/1521

[50] Raben A, Vasilaras TH, Møller AC, Astrup A. Sucrose compared with artificial sweeteners: different effects on ad libitum food intake and body weight after 10 wk of supplementation in overweight subjects. The American Journal of Clinical Nutrition. 2002;**76**:721-729.

[51] Barrett L, Geddes JE, Mangano SF, Schmick F, Whitehouse AS. Chewy Confectionery Product. United States Patent. US 6,531,174. March 11, 2003

[52] Tandel KR. Sugar substitutes: Health controversy over perceived benefits. Journal of Pharmacology and Pharmacotherapeutics. 2011;**2**(4):236-243. DOI:10.4103/0976-500X.85936

[53] Wikipedia: The free encyclopedia. Sugar substitute [Internet]. 2017. Available from: https://en.wikipedia.org/wiki/Sugar_substitute [Accessed: June 26, 2017]

[54] Seventy-sixth Meeting of the Joint FAO/WHO Expert Committee on Food Additives (JECFA). Safety Evaluation of Certain Food Additives. Geneva: World Health Organization, 2012. WHO Food additives series: 67. 978 92 4 166067 9/0300-0923

[55] European Food Additives Commission [Internet]. 2008. Available from: https://webgate.ec.europa.eu/foods_system/main/?sector=FAD [Accessed: May 28, 2017]

[56] Food Standards Agency. Food Additives Legislation: Guidance Notes. London WC2 6NH: 2002

[57] U.S. Food and Drug Administration. Food additives and ingredients [Internet]. 2015. Available from: https://www.fda.gov/food/ingredientspackaginglabeling/foodadditivesingredients/default.htm [Accessed: May 28, 2017]

[58] Wikipedia: The free encyclopedia. Acceptable Daily Intake. [Internet]. 2016. Available from: https://en.wikipedia.org/wiki/Acceptable_daily_intake [Accessed: June 19, 2017]

[59] Notter J. What is innovation? [Internet]. Available from: http://jamienotter.com/2012/09/what-is-innovation/ [Accessed: June 26, 2017]

Section 2

Additives from Plant Origin

Chapter 2

Nutritional, Bioactive and Physicochemical Characteristics of Different Beetroot Formulations

Diego dos S. Baião, Davi V.T. da Silva,
Eduardo M. Del Aguila and
Vânia M. Flosi Paschoalin

Additional information is available at the end of the chapter

http://dx.doi.org/10.5772/intechopen.69301

Abstract

Beetroot possesses high nutritional value and is considered one of the main dietary sources of nitrate. Nitrate has increasingly attracted the interest of the scientific community regarding new physiological, nutritional and therapeutic approaches with beneficial effects on the cardiovascular system. These effects can be explained by the possible effect of dietary nitrate in stimulating nitric oxide synthesis. Dietary nitrate can be reduced to nitrite in the oral cavity, which is then decomposed to nitric oxide and other bioactive nitrogen oxides in the stomach. Beetroot administration can be conducted by several types of formulations, in order to provide a convenient and alternative source of dietary beetroot, such as beetroot juice or beetroot chips and powder. The challenge in providing a product which, in addition to being rich in nitrate, is attractive and easy to administer, while also being microbiologically safe, is increased by the limited scientific information available concerning the nutritional aspects of beetroot formulations. In this chapter, a brief review on the efficiency of different beetroot formulations on health indicators is conducted, emphasizing the effects following the intake of nitrate-enriched beetroot gel. The metabolic and hemodynamic effects of beetroot formulations in healthy and non-healthy volunteers are also discussed.

Keywords: beetroot formulations, nitrate, nitric oxide, phenolic compounds

1. Introduction

Lifestyle and inadequate eating habits expose humans to a number of risk factors for the development of chronic non-communicable diseases (CNCDs). Diets rich in saturated, *trans*-fats and simple sugars, poor in complex-carbohydrates and fibres and associated with smoking, alcoholism, stress and sedentary lifestyles have increased the number of diseases such as obesity, diabetes, hypertension, osteoporosis and cardiovascular disorders, among others [1]. Therefore, the search for a healthy diet has been significantly emphasized worldwide [2].

Vegetables are important components of a balanced diet due to their constituents, mainly bioactive compounds, fibres, vitamins and minerals. Epidemiological studies have shown that vegetables are useful protective foods against coronary heart disease and ischemia [3]. These and other findings have raised the hypothesis increasingly recognized and biologically plausible, that inorganic nitrate (NO_3^-) in certain vegetables can provide a physiological substrate for the production of nitric oxide (NO) and other products (NOx), which in turn cause vasodilation, decrease blood pressure and support cardiovascular function [4].

Dietary NO_3^- can be reduced to nitrite (NO_2^-) in the oral cavity by commensal bacteria that express the enzyme nitrate reductase [4]. NO_2^- then reaches the stomach where, in contact with gastric acid, it is non-enzymatically decomposed into NO and other bioactive nitrogen oxides through the NO_3^--NO_2^-/NO pathway [5]. The theory that dietary NO_3^- and NO_2^- may stimulate the endogenous synthesis of NO has increasingly been noted, and a new physiological, therapeutic and nutritional approach for these anions has arisen [6].

Dietary administration of beetroot, a vegetable NO_3^- source, can be conducted through several types of formulations, in order to provide a convenient and alternative source of beetroot, instead of consuming the whole vegetable. Some beetroot formulations must be offered in large amounts to reach effective NO_3^- concentrations, making it difficult to convince individuals to adhere to certain proposed interventions [7–9]. In this context, this paper discusses the recent advances in beetroot administration, pointing out plant species, nutritional composition and the effect of the ingestion of different beetroot formulations on NO production and the ensuing effects on hemodynamic parameters.

2. Beetroot cultivation

The Beetroot species *Beta vulgaris* L. belongs to the Quenopodiaceae family and originated in regions of Europe and North Africa, where they are cultivated in mild to cold temperatures (10–20°C). Cultivation in climatic conditions with higher relative humidity and higher temperature favours the development of pests and diseases, altering the internal colour and taste of the plant, making it less sweet, also reducing plant productivity by about 50% [10]. This plant species prefers soils rich in organic matter, with pH ranging from 5.5 to 6.2. The production cycle can range from 60 to 100 days, in summer or winter, depending on the cultivar and cultivation mode [10, 11].

The plant has a root system composed of a main root and smaller roots reaching up to 60 cm in depth, with lateral branching. It also possesses a tuberous, purplish-red, part, globular in shape, with a sweet taste, which develops almost on the surface of the soil [12]. The beetroot plant is biennial, requiring a period of intense cold to go through the reproductive stage of the cycle. The appearance of elongated leaves around the stem and the tuberous part occurs in the vegetative phase, while floral tassel emission occurs with the production of seeds comprised of glomeruli during the reproductive stage [11].

According to some authors, *Beta vulgaris* L. beetroots can be divided into three subspecies: (a) *Beta vulgaris* ssp. *adanesis*, formed by a distinct group of semi-annual plants, with a great decline in auto-fertilization, with specific morphological characteristics; (b) *Beta vulgaris* ssp. *maritima*, formed by a large complex of morphological types that occur in a vast geographic area; (c) *Beta vulgaris* ssp. *vulgaris*, which groups all domesticated cultivars [13].

According to Lange et al. [13], subspecies *Beta vulgaris* ssp. *Vulgaris* cultivars can be subdivided into four other groups: (1) **Leaf Beet Group**, a cultivar with edible leaves and petioles and with roots with no significantly increased diameter; (2) **Sugar Beet Group**, a white coloured strain grown in the US and Europe for sugar production; (3) **Fodder Beet Group**, a cultivar intended for feeding herds and (4) **Garden Beet Group**, the only group cultivated in Brazil that has an edible tuberous part.

3. Nutritional composition of beetroot (*Beta vulgaris* L.)

The beetroot species *Beta vulgaris* L. is considered a good source of dietary fibre, minerals (potassium, sodium, iron, copper, magnesium, calcium, phosphorus and zinc), vitamins (retinol, ascorbic acid and B-complex), antioxidants, betalains and phenolic compounds, and possesses high nutritional value due to its high glucose content, in the form of sucrose [5, 14, 15]. According to data presented by the United States Department of Agriculture (USDA) for macronutrients, 100 g of raw beetroot has an energy value of 43 kcal, 9.56 g of carbohydrates, 1.61 g of proteins, 0.17 g of total lipids, 2.8 g of total dietary fibre and 6.76 g of total sugars [15].

4. Bioactive beetroot compounds

Plants are generally considered important sources of substances that perform bioactive functions, favouring human health, good organ function, disease control and contributing to longevity [16–19]. Insufficient intake of bioactive compounds from plant sources is considered an important risk factor for the development of chronic and non-communicable diseases.

Bioactive compounds may have different physiological targets and mechanisms of action, Many of these compounds show antioxidant action due to their potential for oxi-reduction of certain molecules, while others have the capacity to compete for active enzymatic and receptor sites in various subcellular structures or may modulate the expression of genes encoding proteins involved in intracellular mechanisms of defence against oxi-degenerative processes of molecules and cellular structures [20].

4.1. Betalains

Beetroots are a major source of betalains, classified as one of the 10 plants with the highest antioxidant activity, determined by the IC50 value, i.e. the extract's ability to reduce low-density lipoproteins (LDL) oxidation by 50%, as reported by Vinson et al. [21]. Betalains are present in the tuberous part of the plant, conferring its red-purple coloration, and act as antioxidant agents [12]. In addition, anti-inflammatory, hepatic protective and anti-cancer properties have also been attributed to this class of compounds [22, 23].

Betalains are heterocyclic compounds and water-soluble nitrogen pigments responsible for conferring various types of coloration in flowers, vegetables and fruits [24]. Betanin, a compound belonging to one of the etalain classes, has excellent stability at pH 4 and 5 and reasonable stability between pH 3 and 4 and pH 5 and 7. The stability of these molecules is affected by factors such as exposure to light, high temperature (i.e. cooking processes), high water activity and the presence of oxygen [23–27]. In addition, they can also be degraded by enzymes, such as polyphenoloxidases and peroxidases, released during food processing [28, 29]. Betanins are formed from a common precursor, betalamic acid, with more than 50 structures identified so far. Furthermore, these pigments are present in acid form due to the presence of several carboxyl groups and, are, therefore not classified as alkaloids [30].

Betalains have a general structure that contains betalamic acid, accompanied by a radical R1 or R2, where the substituents may be simple hydrogen or a more complex radical. The variability of the substituent groups comes from the different sources of these pigments and determines their hue and stability. According to their chemical structure, betalains are divided into two subclasses: betaxanthins, responsible for orange-yellow colorations, such as vulgaxanthin I and II and indicaxanthin, and betacyanins, responsible for red-violet colorations, such as betanin, prebetanin, isobetanine, neobetanin, amaranthine, gomphrenin and bouganvillein [12, 24] (**Figure 1**).

Previous studies have shown that betacyanins and betaxanthins are capable of absorbing visible light [31–33], and the structural differences between these compounds reflect on their

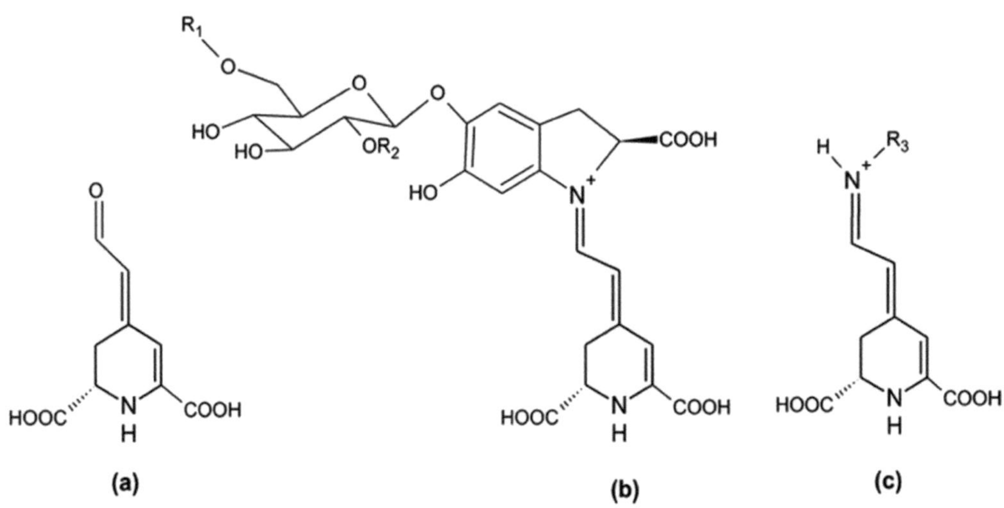

Figure 1: General structures of betalamic acid (a), betacyanins (b) and betaxanthins (c). Betanin: R1 = R2 = H. R3 = amine or amino acid group. *Reproduced from Azeredo et al. [31].

light absorbing capacity. Approximately 50 types of betacyanins and 20 types of betaxanthins have been described. Beetroots contain both, approximately 75–95% betacyanin and 5–25% betaxanthine [24, 33]. The content of these pigments is present in higher concentrations on the vegetable skin, gradually decreasing towards the inner parts of the beet (pulp and crown) [25, 33, 34].

Betalains are widely used in the food industry as natural food dyes, and have shown *in vitro* and *in vivo* antioxidant capability, attributed mostly to betacyanins [33, 35, 36]. Previous studies have demonstrated that low concentrations of the betanin pigment, a beta-cyanine-like compound, present in beetroots, were able to inhibit *in vitro* oxidation of low-density lipoproteins (LDL) by H_2O_2-activatedmetamioglobin, as well as oxidizing agents, such as lipoxygenases and cytochrome 'c' [37]. Motivated by this evidence, several studies have evaluated the intake of foods rich in this pigment for the prevention and treatment of diseases triggered by oxidative processes in humans [31, 37, 38]. For example, Cai et al. [35] observed that betalains obtained from plants belonging to the Amaranthaceae family presented higher antioxidant activity than that of ascorbic acid, a traditional antioxidant.

Several other studies have demonstrated the role of betalains in the protection of several cellular components against oxidative stresses [38–40]. For instance, Reddy et al. [41] evaluated the efficiencies of betanin, anthocyanin, lycopene, bixin, b-carotene and chlorophyll separately and combined in inhibiting lipid peroxidation, cyclooxygenase enzymes and in the proliferation of human tumour cells. The authors observed that among the tested pigments, betanin, lycopene and β-carotene were more efficient at inhibiting lipid peroxidation.

The antioxidant potential of betalains has been attributed to the molecular structure of these compounds, which reflect their ability to donate hydrogen to reactive species. Regarding betaxanthins, an increase in the number of hydroxyl and imino residues promotes the elimination of free radicals, while glycosylation reduced activity and acylation increased antioxidant potential in betacyanins [34]. The antioxidant activity of betalains can be increased according to the number and position of the amino and hydroxyl groups in the molecule, with the C-5 position of the hydroxyl group in the aglycone responsible for increasing their antioxidant activity [31, 35].

Beetroots are the main commercial source of betalains (concentrate or powder) in the food industry, with the use of betanin restricted as a natural dye added to different foods like gelatins, desserts in general, confectionery, dry mixes, dairy products and beef and chicken-derived products [33]. Betalains are unstable and are degraded in the presence of light and oxygen and destroyed at high temperatures. The daily intake limit of betalains has not been established. In Brazil, current legislation dictates that the use of natural red dye obtained from beets is permitted in foods and beverages [33].

4.2. Phenolic compounds

Phenolic compounds present in plants are derived from the aromatic amino acid phenylalanine and are formed by two aromatic rings (A and B), linked by three carbon atoms that form an oxygenated heterocycle (ring C) [42, 43] (**Figure 2**). These compounds are secondary

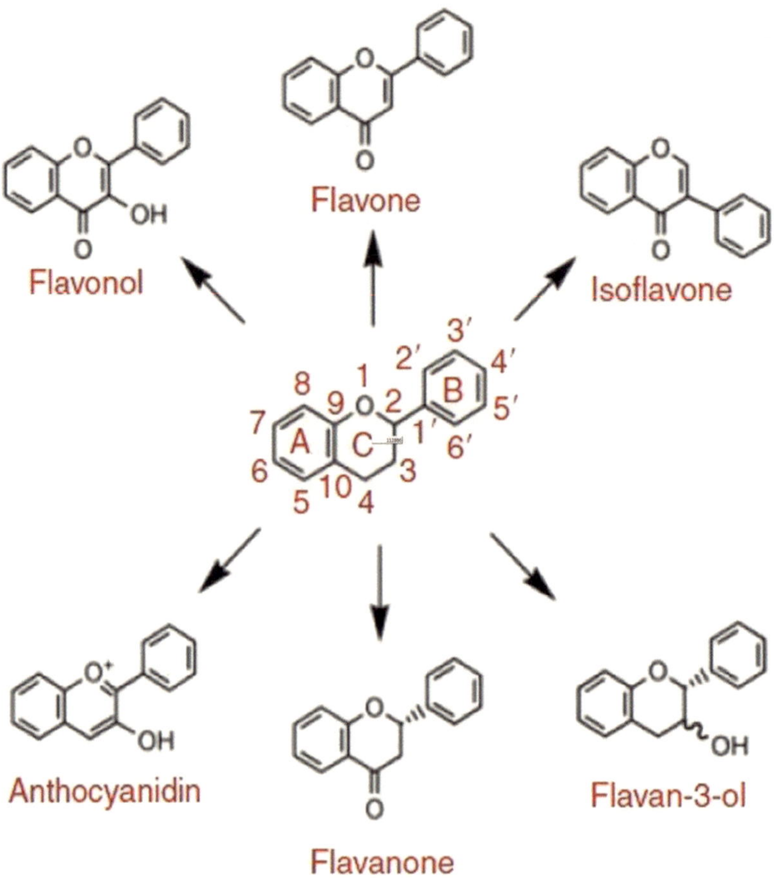

Figure 2: Chemical structure of the major flavonoids. *Reproduced from Crozier et al. [43].

metabolites in fruits and vegetables and are generally involved in the defence against pathogens and/or against ultraviolet radiation. In addition, many volatile phenolic compounds are responsible for the assignment of sensory characteristics, or flavour, in various foods commonly included in the human diet [44].

Beetroots are also an adequate source of phenolic compounds, and several studies have been carried out on the identification and evaluation of the antioxidant content and capacity of phenolic compounds present in this vegetable [45]. Previous studies have shown that beetroot tuber juice is particularly rich in phenolic compounds and that these components remain active even after *in vitro* digestion trials [45, 46]. Kujala et al. [25, 36] reported the presence of ferulic acid, flavonoids and phenolic amides, mainly in beetroot skin, which confer strong antioxidant activity to this food.

Flavonoid concentrations and species present in beetroot formulations should be better evaluated so that the intake of this vegetable in the human diet is as efficient as possible, since flavonoids can be lost in vegetable processing. In a recent study, Silva et al. [9] developed a beetroot gel formulation, obtained from beetroot juice and powder, enriched in flavonoid content. While the beetroot juice contained 0.42 mg of quercetin equivalents per gram (QE/g), phenolic compounds in the gel were three times higher, reaching 1.37 mg QE/g. Guldiken et al. [47] determined flavonoid concentrations in beetroots processed through different forms of cooking, such as oven drying, canning, puree, processed juice, jelly and fresh beetroots, and found values ranging from 1.26 to 2.61 mg of rutin equivalents (RE)/g. Kazimierczak et al. [48] reported average flavonoids values of 0.41–1.16 mg/g in beetroots obtained by conventional and organic planting, respectively.

Phenolic compound composition and concentrations also vary according to the cultivar. Kujala et al. [36] studied four distinct beetroot cultivars by high-performance liquid chromatography (HPLC) and HPLC- electrospray ionisation-mass spectrometry (HPLC-ESI-MS) and NMR techniques and identified four flavonoids, betagarin, betavulgarin, cochliophilin A and dihydroisorhamnetin, whereas Georgiev et al. [49] identified the presence of catechin hydrate, epicatechin and rutin. The importance of flavonoids in the human diet has been ascribed to their horticultural effect, contributing to chemoprevention of DNA damage caused by several carcinogenic factors [49].

4.3. Saponins

These compounds vary extensively in structure and are widely distributed in plants. They are triterpene glycosides wherein the aglycone is covalently linked to one or two sugar chains through a glycosidic ester (C-28) or ether (to C-3) bond. Saponins exhibit several biological activities, such as anti-viral, anti-diabetic and anti-haemolytic properties, and have been, therefore, well studied [50].

Few studies are available regarding the evaluation of saponin content in beetroots. The occurrence of saponins in these species was characterized in a study conducted by Mroczek [51], where 11 saponins derived from oleanoic acids were identified using reverse phase liquid chromatography coupled with electrospray ionization mass spectrometry (LC-ESI/MS/MS). As with flavonoids, the content and species of saponins may vary according to the cultivar, with concentrations varying from 7.66 to 12.2 mg/g dry weight in three different beetroot cultivars (*Beta vulgaris* L.). Oleanoic acid is of importance among the saponins identified in the beet, since it is capable of causing a marked hypoglycemic effect [52]. Saponin content may also vary according to beetroot processing. In the study conducted by Silva et al. [9], saponin content was almost three times higher in beetroot gel compared to juice, of 22 and 8.22 mg/g, respectively.

4.4. Dietary NO_3^- and NO_2^-

NO_3^- and NO_2^- are compounds formed by a single nitrogen bonded to three or two oxygen atoms, respectively. NO_3^- is a nitric acid salt, while NO_2^- is a nitrous acid salt, and both can be obtained from endogenous and/or exogenous sources. NO_3^- are relatively inert but are transformed into NO_2^- by bacteria in the mouth of enzymatic pathways in the body. NO_2^- can be metabolically converted into NO via the L-arginine/NO pathway. This pathway was discovered in 1916 by Mitchell et al. [53] and confirmed by Green et al. [54] and Leaf et al. [55]. In addition, increased NO_2^--plasma concentrations have been derived from this pathway, with no contribution from dietary NO_3^- intake. Vasconcellos et al. [7] and Baião et al. [8] confirmed this assertion by supplementing healthy and physically active volunteers with juice and beetroot gel, respectively. The authors observed an increase in urinary NO_2^- concentrations after juice and beetroot gel consumption, but no change was observed after consumption of a placebo (PLA, juice and beetroot gel with reduced NO_3^- content).

After ingestion, dietary NO_3^- is well absorbed in the upper gastrointestinal tract. About 25% of NO_3^- absorbed via the dietary route is captured by the salivary glands. In the oral cavity, NO_3^- is reduced to NO_2^- by the enzyme NO_3^--reductase. Upon being swallowed, NO_2^- formed from dietary NO_3^- is decomposed non-enzymatically into NO, which is then rapidly oxidized to NO_2^- [56]. Thus, an increase in urinary NO_2^- concentrations after administration of beetroot juice or gel but not after ingestion of a PLA is expected. For this reason, many studies have used urinary or plasma NO_2^- as the main marker for NO production.

The major dietary intake of NO_3^- and NO_2^- by the occidental population comes from vegetables (approximately 85%). The content of these anions in plants vary according to the vegetable tissues. For example, NO_3^- in plant organs can be classified from highest to lowest, as petiole > leaf > stem > root > tuber > bulb > fruit > seed [57]. In addition, environmental factors (atmospheric humidity, temperature, water content and exposure to sunlight and irradiation), agricultural factors (type of crop, fertilization, soil conditions, herbicide use, amount of available nitrogen, availability of other nutrients, plant genotype and transport and storage conditions) also influence NO_3^- levels in plants [58]. NO_2^-, on the other hand, is not found naturally in fruits and vegetables; it is unstable and rapidly oxidized to NO_3^-. However, human exposure to NO_2^- is about 70–80% derived from the additives used during food processing, such as meats, bakery products and cereals to improve taste and appearance, and to prevent the growth of foodborne pathogens and the secretion of toxins, such as the botulin toxin [59].

5. NO

NO is an endogenously produced molecule in the form of a gas with a very short half-life (5–10 s), low molecular weight (30.01 g/mol) and in standard temperature and pressure, moderate solubility in water (1.9 mM, at 25°C). It has a better solubility in polar solvents and in biological systems, and is more concentrated in lipophilic environments (cell membranes and hydrophobic domains of proteins) [60, 61].

The physiological importance of NO is due to the fact that it exerts a second-messenger function, activating or inhibiting several molecules, regulating vascular tone, and acting as an effector of the immune system and neurotransmission. Unlike other intracellular messengers, NO depends on redox reactivity to associate with a receptor or enzyme, not on its molecular structure. In addition, NO is not stored *in vivo* as other neurotransmitters but rather is synthesized on demand and rapidly diffuses into the target tissue and easily and quickly penetrates into other cells due to its small size and lipophilic characteristics [4, 61].

NO has several health benefits. It participates in the regulation of vascular tonus and interacts with the iron of the prosthetic heme group of the soluble guanylate cyclase enzyme (GCs), activating the production of cyclic guanosine monophosphate (cGMP) and consequent relaxation of adjacent smooth muscle cells [62]. In the immune system, it shows cytotoxic and cytostatic effects promoting the destruction of microorganisms, parasites and tumour cells [63]. This compound can also act as a neurotransmitter in the central and peripheral nervous system, facilitating the release of other neurotransmitters and hormones [64]. Other studies have reported that NO may also act on circulating blood cells (monocytes and platelets) for the maintenance of vascular homeostasis and control of smooth muscle cell proliferation and growth, as well as activation and aggregation of platelets, leukocytes and adhesion molecules present in the inflammatory process [65].

6. Production of different beetroot formulations

In the last few years, aiming at obtaining a convenient and alternative source of dietary NO_3^-, different beetroot formulations have been tested, with different nutritional compositions, intending to promote beneficial health effects (**Figure 3**). The challenge is to provide a product that, besides being rich in NO_3^-, is attractive, easy to administer and microbiologically safe.

The first formulation described in several studies for dietary NO_3^- supplementation was beetroot juice [8, 66–73]. For example, Baião et al. [8] produced beetroot juice by thoroughly washing the vegetables in tap water, sanitizing them in a chlorine solution (0.5%) as recommended by the Brazilian Health Ministry legislation ANVISA (MS, Resolution RDC No. 216 of 15/09/2004) and preparing them a food centrifuge processor. The beetroots were processed without adding any additional water.

Other authors have used water, fruit juice or other substances with low NO_3^- content to offer the PLA solution to volunteers [66, 67, 72, 73]. To obtain free-NO_3^- beetroot juice, Baião et al. [9] took the juice prepared in food processor centrifuge and transferred it to a sterile bottle containing an NO_3^- specific anion–exchange resin. After 1 h, the juice was loaded into a sterile glass column and eluted with the aid of a vacuum pump. The free-NO_3^- beetroot juice was similar in colour, taste, appearance and texture to the original. Both enriched and free-NO_3^- beetroot juices were administered to volunteers in order to achieve the health benefits.

Kaimainen et al. [74] and Vasconcellos et al. [7] prepared beetroot powder after freeze-drying beetroot juice in a spray dried system. Vasconcellos et al. [7] also produced beetroot chips. To produce beet powder, beetroots were sanitized in a chlorine solution (0.5%) and after beetroot juice production, the juice was dried at respective inlet and outlet temperatures of 180 and 65 ± 3°C with a 0.7 mm nozzle and a 6 mL/min feed. Beetroot chips were obtained by cutting the beetroot vertically into slices of 3–8 cm wide and 2–4 mm thick, in order to obtain thin round slices. The slices were then frozen at −20°C for 48 h, and subsequently freeze-dried.

Recently, Silva et al. [9] developed a new formulation, a beetroot gel, mainly aimed at the administration of NO_3^- to athletes during sports competitions. The gel was prepared from

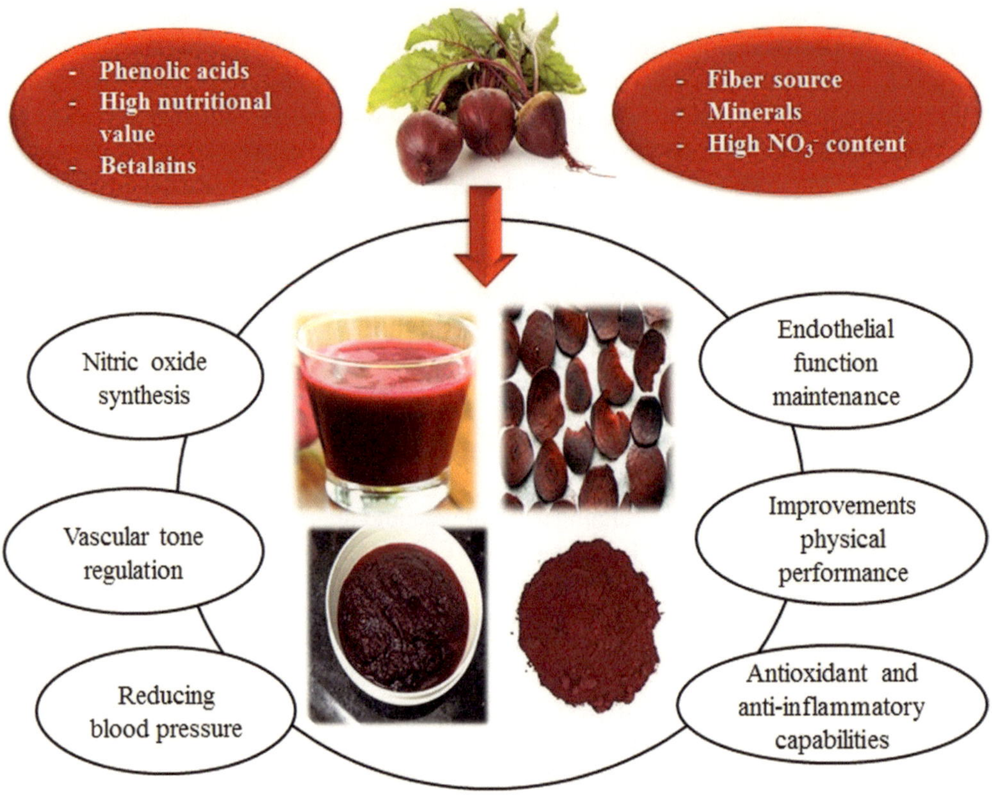

Figure 3: Different beetroot formulations and nutritional benefits.

sanitized vegetables washed in a chlorinated solution (0.5%), half of which was used for juice production, as described previously, and the other half frozen at −20°C for 48 h and then crushed in a portable blender to produce beetroot powder. The beetroot gel was formulated with a mixture of beetroot juice, beetroot powder and carboxymethylcellulose at a 90:17:3 ratio. This new beetroot-based gel presented pseudoplastic fluid characteristics, with decreased apparent viscosity as a function of increasing shear rate (deformation). This pseudoplastic behaviour can be a great advantage for the production and marketing of beetroot gel with respect to handling, packaging and yield, since the product flows smoothly, resulting in a suitable flow from the sachet into the mouth during gel ingestion. According to Rao [75], food rheology has an influence on taste and flavour perception, because of an effect related to the physical properties of the food (i.e. texture, viscosity), which affects the rate and extent to which the stimuli reach the taste buds. Colour is one of the most important food attributes, and is considered a quality indicator. The beetroot gel showed a low lightness, according to the CIE system scale ranging from 0 to 100. Visually, the produced gel had a low brightness (dark) red–violet colour, characteristic of raw beetroots, due to the characteristic colour of betanin, a water-soluble compound derived from betalamic acid. About 101 untrained panelists comprising both males and females (using a 9-point hedonic scale: where one reflected extreme dislike and nine reflected the highest acceptability) evaluated the beetroot gel. The beetroot gel with orange flavouring received higher mean scores in all sensory attributes (flavour, aroma, texture, and overall impression). It is possible that orange flavouring positively influenced the appraisal of the texture and overall impression attributes by the evaluators. Thus, the NO_3^--enriched beetroot gel presented advantageous rheological properties for oral administration and handling in the nutritional supplement industry. This encourages the testing of beetroot gel in larger groups, such as physically active subjects.

In their most recent study, Vasconcellos et al. [76] produced a beetroot gel with reduced NO_3^- content as previously described by Silva et al. [9] and supplemented 25 healthy runners, and also produced a PLA gel. The PLA was prepared by depleting the NO_3^- from the beetroot juice by using an A-520E anion-exchange resin. Fuji apple (*Malus pumila* species) puree was prepared by liquefying apples in a portable blender. NO_3^--depleted beetroot juice, 16.8 g of apple puree (in substitution of beet chips), 2.8 g of carboxymethylcellulose and 1 mL of artificial orange flavour were then mixed to produce PLA gel.

Table 1 displays the centesimal composition and sugar content of the different beetroot-based formulations in 100 g. Beetroot gel presented the highest protein, lipid and total dietary fiber contents when compared to the other formulations. It is important to note that lipid amount was extremely low in all formulations. Beetroot chips had the highest energy content (in the form of kcal), carbohydrates and total sugars when compared to beetroot juice and gel.

Although beetroot is considered one of the main sources for the acquisition of dietary NO_3^- and of functional compounds with antioxidant activity, the contents of these substances vary significantly from one formulation to another. **Table 2** displays the antioxidant potential, total phenolic compounds content, flavonoids, saponins, NO_3^- and NO_2^- contents of different beetroot formulations. Beetroot gel presented the highest total phenolic content (TPC), flavonoids, saponins and NO_3^- content when compared to juice and chips. However, beetroot chips presented the highest values of the total antioxidant potential (TAP) when compared to the other formulations. Beetroot juice showed the lowest NO_3^- concentrations, while chips showed the lowest TPC, flavonoids and saponin levels when compared to the other formulations, indicating that beetroot processing leads to impoverishment of functional compounds. The NO_2^- content of all formulations was considered insignificant (<1 mmol·100 g).

Although beetroot juice has become the most commonly used formulation for the dietary administration of NO_3^-, beetroot gel showed significantly higher levels of $NO_3^-$and bioactive compounds. Thus, beetroot gel seems to be the most effective formulation, providing a ready, easy to administer, attractive, NO_3^--rich food with the aim of promoting beneficial effects on the cardiovascular system.

Beetroot formulations	kcal	CHO (g)	PTN (g)	LIP (g)	TDF (g)	Total sugars (g)	Fructose (g)	Glucose (g)	Sucrose (g)	Moisture (%)	Ashe (%)
Raw	43	9.56	1.61	0.17	2.8	6.76	-	-	-	87.58	-
Juice	94.9 ± 1.7^{b}	22.6 ± 0.4^{b}	0.70 ± 0.07^{b}	0.16 ± 0.01^{b}	0.91 ± 0.31^{c}	12.1 ± 0.1^{b}	0.86 ± 0.01^{b}	2.5 ± 0.02^{a}	8.8 ± 0.03^{b}	76.1 ± 0.50^{a}	0.8 ± 0.06^{b}
Chips	365.0 ± 2.1^{a}	89.9 ± 0.52^{a}	0.97 ± 0.01^{b}	0.14 ± 0.01^{b}	3.2 ± 0.63^{b}	18.8 ± 0.01^{a}	1.4 ± 0.01^{a}	2.7 ± 0.01^{a}	14.6 ± 0.01^{a}	4.6 ± 0.57^{b}	1.0 ± 0.05^{b}
Gel	71.52 ± 1.9^{c}	13.6 ± 0.31^{c}	3.02 ± 0.09^{a}	0.56 ± 0.01^{a}	4.5 ± 0.28 ^{a}	9.7 ± 0.07^{c}	0.31 ± 0.01^{c}	1. ± 0.01^{b}	8.1 ± 0.05^{b}	76.14 ± 0.5^{a}	2.01 ± 0.13^{a}

The values are displayed as means ±SD. Different letters denote statistical significance between the samples at $P < 0.05$ Kcal, kilocalorie; CHO, carbohydrate; PTN, protein; LIP, lipid; TDF, total dietary fibre; g, gram.

* Reproduced from da Silva et al.[9]; USDA. [15].

Table 1. Proximate composition of different beetroot formulations (100 g).

Beetroot formulations	TAP	TPC	Flavonoids	Saponins	NO_3^-		NO_2^-	
	%	GAE mg	mg	mg	mmol	mg	mmol	mg
Juice	79.13 ± 0.63[c]	1.01 ± 0.03[b]	0.42 ± 0.01[b]	8.22 ± 0.12[b]	1.6 ± 0.01[c]	217[c]	0.10 ± 0.02[c]	4.6[c]
Chips	95.70 ± 0.53[a]	0.75 ± 0.06[b]	0.31 ± 0.02[b]	6.37 ± 1.26[c]	4.5 ± 0.02[b]	**279[b]**	0.13 ± 0.02[b]	5.98[b]
Gel	87 ± 0.1[b]	1.98 ± 0.03[a]	1.37 ± 0.03[a]	22 ± 0.54[a]	6.3 ± 0.41[a]	390[a]	0.15 ± 0.0[a]	6.9[a]

The values are displayed as means ±SD. Different letters denote statistical significance between the samples at $P < 0.05$. TAP, total antioxidant potential; TPC, total phenolic content; GAE, gallic acid equivalents; NO_3^-, nitrate content; NO_2^-, nitrite content.

*Reproduced from da Silva et al.[9]; Vasconcellos et al. [7].

Table 2. Antioxidant potential, total phenolic compounds, bioactive compounds, NO_3^- and NO_2^- contents of different beetroots formulation (100 g).

7. Effect of the ingestion of beetroot formulations on nitric oxide production and consequent health benefits

Due to its high NO_3^- content, beetroots have been used as a dietary source of this anion for the production of NO, aiming at blood pressure lowering effects [77]. The first study conducted with beetroot juice (500 mL) was performed by Webb et al. [66]. The authors offered 500 mL of beetroot juice to 14 healthy volunteers (containing ≈ 22.5 mmol of NO_3^-). Following ingestion, significant increases in NO synthesis (plasma NO_3^- and NO_2^-), stabilization of the endothelial function evaluated by the dilation of the brachial artery-mediated flow (DILA) and a significant decrease of the systolic blood pressure (SBP) by up to 10 mmHg and diastolic blood pressure (DBP) by up to 8 mmHg were observed. In addition, the decreases in SBP and DBP were correlated with increased NO synthesis. Subsequently, the juice was used in other studies that, in addition to healthy subjects, involved individuals with hypertension and associated morbidities [67, 71, 78]. Kapil et al. [78] supplemented 68 hypertensive subjects with 250 mL of beetroot juice (containing ≈ 6.4 mmol of NO_3^-) for 4 weeks and evaluated endothelial function through DILA, plasma NO_3^-, NO_2^-, cGMP (other marker of NO synthesis) and monitored BP before and after the interventions. The authors observed an improvement in endothelial function through a significant increase in the mediated flow of the brachial artery and the concentration of NO_3^- and NO_2^- plasma after 4 weeks of beetroot juice ingestion. Significant SBP decreases by 7.7 mmHg and DBP by 2.5 mmHg were also observed after dietary NO_3^- intake.

Baião et al. [8] acutely supplemented 40 healthy volunteers (20 men and women) with 1.6 mmol of NO_3^- with100 mL beetroot juice. Significant increases in urinary NO_3^-, NO_2^- and NOx (NO_3^- + NO_2^-) concentrations after the ingestion of this small amount of beetroot juice were found, but no differences in the NO metabolites excretion responses between men and women. Regardless of gender and body mass, urinary excretion of NO metabolites after consumption of a dietary source of NO_3^- increased in both men and women.

Table 3 shows studies that evaluated the effect of NO_3^- supplementation through different beetroot formulations regarding their efficiency on NO production and the effect

Study	Beetroot formulations	NO_3– concentration	Experimental population	Duration of administration (days)	Effect
Webb et al. [66]	Juice (500 mL)	22.5 mmol	14 healthy males	Acute	Increases NO_3^- (≈16-fold) and NO_2^- (≈2-fold) in plasma. Decreases of 10.4 + 3 mmHg in SBP and after 24 h (−6 mmHg), decrease of8.1 + 2.1 mmHg in DBP.
Kapil et al. [67]	Juice (250 mL)	5.5 mmol	9 healthy subjects	Acute	Increases NO_3^- (≈2-fold) and NO_2^- (≈1.6-fold) in plasma. Decreases of 5.4 ± 1.5 mm Hg in SBP. No effect on DBP.
Kenjale et al. [79]	Juice (500 mL)	9.0 mmol	8 elderly subjects	Acute	Increases plasma NO metabolites (NO_3^- and NO_2^-). There was an increase in 32 s in exercise time after beetroot juice consumption before subjects reported pain due to claudication. There was a reduction 48% after beetroot juice consumption (indicating that oxygen extraction was reduced) compared to PLA after exercise.
Coles and Clifton [72]	Juice + Apple (500 g)	15.0 mmol/L	30 healthy subjects	Acute	There was a non-significant reduction in SBP and DBP in men and women after beetroot juice consumption when compared to PLA.
Hobbs et al. [69]	Juice (100 mL) (250 mL) (500 mL)	2.3 mmol 5.7 mmol 11.4 mmol	18 healthy males	Acute	Increase in urinary NO. Dose-dependent reduction with peaks of 13.1, 20.5 and 22.2 mmHg in SBP and 16.6, 14.6 and 18.3 mmHg in DBP at doses of 100, 250 and 500 mL, respectively.
Hobbs et al. [69]	Red and white beetroot bread (100 g)	1.6 mmol and 1.8 mmol, respectively.	14 healthy males	Acute	Peak differences in SBP and DBP in the order of 19.3 and 23.6 mmHg and 16.5 and 23.2 mmHg for breads enriched with white and red beetroot, respectively, in relation to the control.
Hobbs et al. (2013) [80]	Red beetroot bread (100 g)	1.1 mmol	24 healthy males	Acute	Increases NO_3^- (≈3-fold) and NO_2^- (≈1-fold) in plasma. Decreases peaks of 7 mmHg in SBP. No effect on DBP.
Gilchrist et al. [71]	Juice (250 mL)	7.5 mmol	27 hypertensive and diabetic type 2	15	Increases NO_3^- (≈5-fold) and NO_2^- (≈1.7-fold) in plasma. No effect on SBP and DBP
Bond Jr et al. [81]	Juice (500 mL)	12.1 mmol	12 healthy women	Acute	Increase plasma NO metabolite (NO_2^-). Maximum decrease of 5.0, 8.1, 6.5, 11.2 mmHg and 3.6, 2.1, 0.4, 3.1 mmHg on SBP and DBP at rest and 40, 60 and 80% VO_{2peak} in submaximal exercise, respectively.

Study	Beetroot formulations	NO_3^- concentration	Experimental population	Duration of administration (days)	Effect
Kim et al. [82]	Juice (140 mL)	12.9 mmol	12 healthy men	Acute	Increases plasma NO metabolites (NO_3^- and NO_2^-). There was no decrease in SBP and DBP after beetroot juice ingestion. There was a significant reduction in PWV 3 h after beetroot juice consumption I was observed an increase of diameter of the brachial artery (mm) and FMD (%) through exercise intensity after beetroot and PLA juice consumption (there was no difference between the interventions).
Jajja et al. [83]	Juice (70 mL)	6.45 mmol	24 overweight subjects	21	Increases NO_3^- in plasma and urine. There was no difference after the interventions beetroot and PLA juice in SBP and DBP.
Bondonno et al. [84]	Juice (140 mL)	7 mmol	27 hypertensive treated patients	7	Increases NO_3^- (≈3-fold) and NO_2^- (≈3-fold) in plasma. No effect on SBP and DBP
Kapil et al. [78]	Juice (250 mL)	6.4 mmol	34 hypertensive treated patients and 34 with hypertension	28	Increases NO_3^- (≈5.5-fold) and NO_2^- (≈2.7-fold) in plasma. Maximum decrease of 8.1 mmHg and 3.8 mmHg on SBP e DBP, respectively.
da Silva et al. [9]	Gel (100 g)	6.3 mmol	five healthy subjects	Acute	Increases NO_2^- in plasma. Decreases in SBP of 6.2 mmHg and in DBP 5.2 mmHg.
Vasconcellos et al. [76]	Gel (100 g)	10 mmol	25 healthy runners	Acute	Increase NO_2^- in urine. There was no improvement in VO_{2peak} and time to fatigue. SBP and DBP did not differ significantly at any of the investigated time points.

NO_3^-, nitrate; NO_2^-, nitrite; NO, nitric oxide; DBP, diastolic blood pressure; SBP, systolic blood pressure; PLA, placebo; VO_{2peak}, maxim oxygen volume; PWV, pulse wave velocity; FMD, mediated flow dilatation.
Beetroot results from the following references [7, 9, 66, 67, 69, 71, 72, 78–84].

Table 3. Studies evaluating the effect of the ingestion of different forms of beetroot on NO production and blood pressure in healthy and hypertensive individuals.

on blood pressure of healthy and hypertensive subjects. In the study conducted by Silva et al. [9], the NO_3^- and antioxidant-enriched beetroot gel was administered acutely to five healthy subjects. About 60 min after a single-dose intake of 100 g of the beetroot gel containing ≈ 6.3 mmol of NO_3^-, a 3-fold increase in plasma NO_2^- was observed, followed by decreases in SBP and DBP of 6.2 mmHg in 60 min and 5.2 mmHg in 120 min, respectively. In the recent study by Vasconcellos et al. [76], supplementation with NO_3^--enriched beetroot gel (100 g containing ≈ 10 mmol of NO_3^-) in 25 healthy and physically active individuals resulted in a significant increase in urinary NO_3^- and NO_2^-levels after 90 min. The study also demonstrated a decrease in hypoglycaemia, sustained even in the post-recovery period, indicating that other significant metabolic changes appear to occur after NO_3^- intake. In addition, the authors also indicated that physically trained individuals would not benefit from increased levels of NO, since no improvement in physical performance was observed during aerobic submaximal exercise assessed by an aerobic exercise protocol on a treadmill (3 min warm-up of 40% peak oxygen consumption, 4 min to 90% of gas exchange threshold I and 70% (Δ) maximal end speed until volitional fatigue). However, no significant changes were observed in systolic and diastolic pressures or cortisol and lactate levels after the ingestion of the beetroot gel.

8. Conclusions

The effects of beetroot intake on cardiovascular health with regard to NO production and blood pressure are well documented in the literature. The bioactive compounds and *in vitro* antioxidant capacity of the formulations point to the importance of the inclusion of this plant as a dietary component aimed at the prevention of cardiovascular diseases, which may also aid in the cellular response to oxidative stress. The present review highlighted the existence of different beetroot formulations, among which are gels enriched in NO_3^- and bioactive/functional compounds formulated by our research group. Studies involving the effects of different forms of beetroot on a significant number of healthy individuals with cardiovascular problems, as well as *in vivo* effect in animal models, could boost the food industry in the elaboration of beetroot formulations as a food support for the population.

Acknowledgements

The authors would like to acknowledge the financial support from CNPq (ConselhoNacional de DesenvolvimentoCientífico e Tecnológico), CAPES (Coordenação de Aperfeiçoamento de Pessoal de Nível Superior) and FAPERJ (Fundação Carlos ChagasFilho de Amparo à Pesquisa do Estado do Rio de Janeiro).

Author details

Diego dos S. Baião, Davi V.T. da Silva, Eduardo M. Del Aguila and Vânia M. Flosi Paschoalin*

*Address all correspondence to: paschv@iq.ufrj.br

Federal University of Rio de Janeiro, Chemistry Institute, Rio de Janeiro (RJ), Brazil

References

[1] WHO. Food and Agriculture Organization. Diet, Nutrition and the Prevalence of Chronic Diseases. Report ofa Joint WHO/FAO Expert Consultation. Geneva (Technical Report Series 916) [Internet]. 2003. Available from: http://www.who.int/dietphysicalactivity/publications/trs916/download/en/ [Accessed: 13 February 2017]

[2] Hasler CM. The changing face of functional foods. The Journal of the American College of Nutrition. 2000;**19**:499S–506S. DOI: 10.1080/07315724.2000.10718972

[3] Joshipura KJ, Hu FB, Manson JE, Stampfer MJ, Rimm EB, Speizer FE, Colditz G, Ascherio A, Rosner B, Spiegelman D, Willett WC. The effect of fruit and vegetable intake on risk for coronary heart disease. Annals of Internal Medicine. 2001;**134**:1106-1114. DOI: 10.7326/0003-4819-134-12-200106190-00010

[4] Lundberg JO, Gladwin MT, Ahluwalia A, Benjamin N, Bryan NS, Butler A, Cabrales P, Fago A, Feelisch M, Ford PC, Freeman BA, Frenneaux M, Friedman J, Kelm M, Kevil CG, Kim-Shapiro DB, Kozlov AV, Lancaster Jr JR, Lefer DJ, McColl K, McCurry K, Patel RP, Petersson J, Rassaf T, Reutov VP, Richter-Addo GB, Schechter A, Shiva S, Tsuchiya K, van Faassen EE, Webb AJ, Zuckerbraun BS, Zweier JL, Weitzberg E. Nitrate and nitrite in biology, nutrition and therapeutics. Nature Chemical Biology. 2009;**5**:865-869. DOI: 10.1038/nchembio.260

[5] Lundberg JO, Weitzberg E, Gladwin MT. The nitrate-nitrite-nitric oxide pathway in physiology and therapeutics. Nature Reviews Drug Discovery. 2008;**7**:156-167. DOI: doi: 10.1038/nrd2466

[6] Spiegelhalder B, Eisenbrand G, Preussman R. Influence of dietary nitrate on nitrite content of human saliva: Possible relevance to in vivo formation of N-nitroso compounds. Food and Cosmetics Toxicology. 1976;**14**:545-548. DOI: 10.1016/S0015-6264(76)80005-3

[7] Vasconcellos J, Conte-Junior CA, da Silva DVT, Pierucci APTR, Paschoalin VMF, Alvares TS. Comparison of total antioxidant potential, and total phenolic, nitrate, Sugar, and organic acid contents in beetroot juice, chips, powder, and cooked beetroot. Food Science and Biotechnology. 2016;**25**:79-84. DOI: 10.1007/s10068-016-0011-0

[8] Baião DS, Conte-Junior CA, Paschoalin VMF, Alvares TS. Beetroot juice increase nitric oxide metabolites in both men and women regardless of body mass. International Journal of Food Sciences and Nutrition. 2016b;**67**:40-46. DOI: 10.3109/09637486.2015.1121469

[9] da Silva DVT, Silva FO, Perrone D, Pierucci APTR, Conte-Junior CA, Alvares TS, Aguila EMD, Paschoalin VMF. Physicochemical, nutritional, and sensory analyses of a nitrate-enriched beetroot gel and its effects on plasmatic nitric oxide and blood pressure. Journal of Food and Nutrition Research. 2016;**60**:1-9. DOI: 10.3402/fnr.v60.29909

[10] Tullio JA, Otto RF, BoerA, Ohse S. Cultivo de beterraba em ambientes protegidos e natura na época de verão. Revista Brasileira de Engenharia Agrícola e Ambiental. 2013;**17**:1074-1079. DOI: 10.1590/S1415-43662013001000008

[11] Sediyama MAN, Santos MR, Vidigal SM, Salgado LT. Produtividade e exportação de nutritentes em beterraba cultivada com cobertura morta e adubação orgânica. Revista Brasileira de Engenharia Agrícola e Ambiental. 2011;**15**:883-889. DOI: 10.1590/S1415-43662011000900002

[12] Ravichandran K, Saw NMMT, Mohdaly AAA, Gabr AMM, Kastell A, Riedel H, Cai Z, Knorr D, Smetanska I. Impact of processing of red beet on betalain content and antioxidant activity. Food Research International. 2013;**50**:670-675. DOI: 10.1016/j.foodres.2011.07.002

[13] Lange W, Brandenburg WA, Bock TSM. Taxonomy and cultonomy of beet (*Beta vulgaris* L.). Botanical Journal of the Linnean Society. 1999;**130**:81-96. DOI: 10.1006/bojl.1998.0250

[14] van Velzen AG, Sips AJ, Schothorst RC, Lambers AC, Meulenbelt J. The oral bioavailability of nitrate from nitrate-rich vegetables in humans. Toxicology Letters. 2008;**181**:177-181. DOI: 10.1016/j.toxlet.2008.07.019

[15] United States Department of Agriculture. USDA National Nutrient Database for Standard Reference [Internet]. 2013. Available from: http://ndb.nal.usda.gov/ndb/foods/list [10 February 2017]

[16] Sabaté J. The contribution of vegetarian diets to health and disease: A paradigm shift? The American Journal of Clinical Nutrition. 2003;**78**:502S–507S. DOI: PMID: 12936940

[17] Jacobs DR, Tapsell LC. Food, not nutrients, is the fundamental unit in nutrition. Nutrition Reviews. 2007;**65**:439-450. DOI: 10.1111/j.1753-4887.2007.tb00269.x

[18] Holst B, Williamson G. Nutrients and phytochemicals: From bioavailability to bioefficacy beyond antioxidants. Current Opinion in Biotechnology. 2008;**19**:73-82. DOI: 10.1016/j.copbio.2008.03.003

[19] Williamson G, Holst B. Dietary reference intake (DRI) value for dietary polyphenols: Are we heading in the right direction? British Journal of Nutrition. 2008;**99**:55-58. DOI: 10.1017/S0007114508006867

[20] Liu RH. Potential synergy of phytochemicals in cancer prevention: Mechanism of action. Journal of Nutrition. 2004;**134**:3479S–3485S. DOI: PMID: 15570057

[21] Vinson JA, Hao Y, Su X, Zubik L. Phenol antioxidant quantity and quality in foods: Vegetables. Journal of Agricultural and Food Chemistry. 1998;**46**:3630-3634. DOI: 10.1021/jf980295o

[22] Kapadia GJ, Azuine MA, Sridhar R, Okuda Y, Tsuruta A, Ichiishi E, Mukainake T, Takasaki M, Konoshima T, Nishino H, Tokuda H. Chemopreventionof DMBA-induced UV-B promoted, NOR-1-induced TPA promoted skin carcinogenesis, and DEN-induced phenobarbital promoted liver tumors in mice by extract of beetroot. Pharmacological Research. 2003;**47**:141-148. DOI: 10.1016/S1043-6618(02)00285-2

[23] Winkler C, Wirleitner B, Schroecksnadel K, Schennach H, Fuchs D. *In vitro* effects of beet root juice on stimulated and unstimulated peripheral blood mononuclear cells. Applied Biochemistry and Biotechnology. 2005;**1**:180-185. DOI: 10.3844/ajbbsp.2005.180.185

[24] Volp ACP, Renhe IRT, Stringueta PC. Pigmentos naturais bioativos. Alimentos e Nutrição, Araraquara. 2009;**20**:157-166. DOI: ISSN 0103-4235

[25] Kujala TS, Loponen JM, Klika KD, Pihlaja K. Phenolics and betacyanins in red beetroot (*Beta vulgaris*) root: Distribuition and effect of cold storage on the content of total phenolics and three individual compounds. Journal of Agricultural and Food Chemistry. 2000;**48**:5338-5342. DOI: 10.1021/jf000523q

[26] Stintzing FC, Trichterborn J, Carle R. Characterisation of anthocyanin-betalain mixtures for food colouring by chromatic and HPLC-DAD-MS analyses. Food Chemistry. 2006;**94**:296-309. DOI: 10.1016/j.foodchem.2005.01.018

[27] Cai YZ, Sun M, Corke H. Colorant properties and stability of Amaranthus betacyanin pigments. Journal of Agricultural and Food Chemistry. 1998;**46**:4491-4495. DOI: 10.1021/jf980457g

[28] Cai Y, Sun M, Schliemann W, Corke H. Chemical stability and colorant properties of betaxanthin pigments from Celosia argentea. Journal of Agricultural and Food Chemistry. 2001;**49**:4429-4435. DOI: 10.1021/jf0104735

[29] Huang AS, Von Elbe JH. Effect of pH on the degradation and regeneration of betanine. Journal of Food Science. 1987;**52**:1689-1693. DOI: 10.1111/j.1365-2621.1987.tb05907.x

[30] Escribano J, Gandía-Herrero F, Caballero N, Pedreño MA. Subcellular localization and isoenzyme pattern of peroxidase and polyphenol oxidase in beet root (*Beta vulgaris L*). Journal of Agricultural and Food Chemistry. 2002;**50**:6123-6129. DOI: 10.1021/jf020356p

[31] Azeredo HM. Betalains: Properties, sources, applications, and stability – a review. International Journal of Food Science & Technology. 2009;**44**:2365-2376. DOI: 10.1111/j.1365-2621.2007.01668.x

[32] Luíz RC, Hirata TAM, Clemente E. Cinética de inativação da polifenoloxidase e peroxidase de abacate (Persea americana Mill.). Ciencia e Agrotecnologia. 2007;**31**:1766-1773. DOI: 10.1590/S1413-70542007000600025

[33] Delgado-Vargas F, Jiménez AR, Paredes-López O. Natural pigments: Carotenoids, anthocyanins, andbetalains - characteristics, biosynthesis, processing, andstability. Critical Reviews in Food Science and Nutrition. 2000;**40**:173-289. DOI: 10.1080/10408690091189257

[34] Stintzing FC, Carle R. Functional properties of anthocyanins and betalains in plants, food and in human nutrition. Trends in Food Science & Technology. 2004;**15**:19-38. DOI: 10.1016/j.tifs.2003.07.004

[35] Cai Y, Sun M, Corke H. HPLC characterization of betalains from plants in the amaranthaceae. Journal of Chromatographic Science. 2005;**43**:454-460. DOI: 10.1093/chromsci/43.9.454

[36] Kujala TS, Vienola MS, Klika KD, Loponen JM, Pihlaja K. Betalain and phenolic compositions of four beetroot (*Beta vulgaris*) cultivars. European Food Research and Technology. 2002;**214**:505-510. DOI: 10.1007/s00217-001-0478-6

[37] Gandía-Herrero F, Cabanes J, Escribano J, García-Carmona F, Jiménez-Atiénzar M. Encapsulation of the most potent antioxidant betalains in edible matrices as powders of different colors. Journal of Agricultural and Food Chemistry. 2013;**61**:4294-4302. DOI: 10.1021/jf400337g

[38] Kanner J, Harel S, Granit R. Betalains-a new class of dietary cationized antioxidants. Journal of Agricultural and Food Chemistry. 2001;**49**:5178-5185. DOI: 10.1021/jf010456f

[39] Tesoriere L, Allegra M, Butera D, Livrea MA. Absorption, excretion, and distribution of dietary antioxidant betalains in LDLs: Potential health effects of betalainsin humans. American Journal of Clinical Nutrition. 2004;**80**:941-945. DOI: PMID:1544790

[40] Tesoriere L, Fazzari M, Angileri F, Gentile C, Livrea MA. In vitro digestion of betalainic foods. Stability and bioaccessibility of betaxanthins and betacyanins and antioxidative potential of food digesta. Journal of Agricultural and Food Chemistry. 2008;**56**:10487-10492. DOI: 10.1021/jf8017172

[41] Reddy MK, Alexander-Lindo RL, Nair MG. Relative inhibition of lipid peroxidation, cyclooxygenase enzymes, and human tumor cell proliferation by natural food colors. Journal of Agricultural and Food Chemistry. 2005;**56**:9268-9273. DOI: 10.1021/jf051399j

[42] Manach C, Scalbert A, Morand C, Rémésy C, Jiménez L. Polyphenols: Food sources and bioavailability. American Journal of Clinical Nutrition. 2004;**79**:727-47. PMID: 15113710

[43] Crozier A, Clifford MN, Ashihara H. Plant Secondary Metabolites: Occurrence, Structure and Role in the Human Diet. Singapore: Blackwell Publishing; 2006. p. 353. DOI: 10.1002/9780470988558

[44] Aherne SA, O'brien NM. Dietary flavonols: Chemistry, food content, and metabolism. Nutrition. 2002;**18**:75-81. DOI: 10.1016/S0899-9007(01)00695-5

[45] Wootton-Beard PC, Moran A, Ryan L. Stability of the total antioxidant capacity and total polyphenol content of 23 commercially available vegetable juices before and after in vitro digestion measured by FRAP, DPPH, ABTS and Folin–Ciocalteu methods. Food Research International. 2011;**44**;217-224. DOI: 10.1016/j.foodres.2010.10.033

[46] Wootton-Beard PC, Ryan L. A beetroot juice shot is a significant and convenient source of bioaccessible antioxidants. Journal of Functional Foods. 2011;**3**:329-334. DOI: 10.1016/j.jff.2011.05.007

[47] Guldiken B, Toydemir G, Memis KN, Okur S, Boyacioglu D, Capanoglu E. Home-processed red beetroot (*Beta vulgaris* L.) products: Changes in antioxidant properties and bioaccessibility. International Journal of Molecular Sciences. 2016;**17**:858. DOI: 10.3390/ijms17060858

[48] Kazimierczak R, Hallmann E, Lipowski J, Drela N, Kowalik A, Püssa T, Matt D, Luik A, Gozdowski D, Rembiałkowska E. Beetroot (*Beta vulgaris* L.) and naturally fermented beetroot juices from organic and conventional production: Metabolomics, antioxidant levels and anticancer activity. Journal of the Science of Food and Agriculture. 2014;**94**:2618-2629. DOI: 10.1002/jsfa.6722

[49] Georgiev VG, Weber J, Kneschke EM, Denev PN, Bley T, Pavlov AI. Antioxidant activity and phenolic content of betalain extracts from intact plants and hairy root cultures of the red beetroot *Beta vulgaris* cv. Detroit dark red. Plant Foods for Human Nutrition. 2010;**65**:105-111. DOI: 10.1007/s11130-010-0156-6

[50] Mikołajczyk-Batora K, Błaszczyka A, Czyżniejewskib M, Kachlickib P. Characterisation and identification of triterpenesaponins in the roots of red beets (*Beta vulgaris* L.) using two HPLC–MS systems. Food Chemistry. 2016;**192**:979-990. DOI: 10.1016/j.foodchem.2015.07.111

[51] Mroczek A. Triterpenesaponin content in the roots of red beet (*Beta vulgaris L.*) cultivars. Journal of the Science of Food and Agriculture. 2012;**60**:12397-12402. DOI: 10.1021/jf303952x

[52] Murakami T, Matsuda H, Inadzuki M, Hirano K, Yoshikawa M. Medicinal foodstuffs. XVI. Sugar Beet. (3): Absolute stereostructures of betavulgarosides II and IV, hypoglycemic saponins having a unique substituent, from the roots of *Beta vulgaris* L. Chemical and Pharmaceutical Bulletin. 1999;**47**:1717-1724. DOI: 10.1248/cpb.47.1717

[53] Mitchell HH, Shonle HA, Grindley HS. The origin of the nitrates in the urine. Journal of Biological Chemistry. 1916;**24**:461-490. Available from: http://www.jbc.org/content/24/4/461.citation [Accessed: 25-January-2017]

[54] Green LC, Ruiz de Luzuriaga K, Wagner DA, Rand W, Istfan N, Young VR, Tannenbaum SR. Nitrate biosynthesis in man. Proceedings of the National Academy of Science of the United States of America. 1981;**78**:7764-7768. DOI: PMCID: PMC349351

[55] Leaf CD, Wishnok JS, Tannenbaum SR. L-arginine is a precursor for nitrate biosynthesis in humans. Biochemical and Biophysical Research Communications. 1989;**163**:1032-1037. DOI: 10.1016/0006-291X(89)92325-5

[56] Lidder S, Webb AJ. Vascular effects of dietary nitrate (as found in green leafy vegetables and beetroot) via the nitrate-nitrite-nitric oxide pathway. British Journal of Clinical Pharmacology. 2012;**75**:677-696. 10.1111/j.1365-2125.2012.04420.x

[57] Hord NG, Tang Y, Bryan NS. Food sources of nitrates and nitrites: The physiologic context for potential health benefits. American Journal of Clinical Nutrition. 2009;**90**:1-10. DOI: 10.3945/ajcn.2008.27131

[58] Anjana SUIM, Abrol YP. Are nitrate concentrations in leafy vegetables within safe limits? Current Science. 2007;**92**:355-360. DOI: ISSN 0011-3891

[59] Pennington J. Dietary exposure models for nitrates and nitrites. Food Control. 1998;**9**:385-395. DOI: 10.1016/S0956-7135(98)00019-X

[60] Archer S. Measurement of nitric oxide in biological models. FASEB Journal. 1993;**7**:349-353. DOI: 10.1111/bph.12832

[61] Barreto RL, Correia CRD. Óxido Nítrico: Propriedades e Potenciais usos Terapêuticos. Quim Nova. 2005;**28**:1046-1054. DOI: 10.1590/S0100-40422005000600020

[62] Bryan NS, Ivy JL. Inorganic nitrite and nitrate: Evidence to support consideration as dietary nutrients. Nutrition Research. 2015;**35**:643-654. DOI: 10.1016/j.nutres.2015.06.001

[63] James SL. Role of nitric oxide in parasitic infections. Microbiology Reviews. 1995;**59**:533-547. DOI: PMCID: PMC239385

[64] Dawson DA. Nitric oxide and focal cerebral ischemia: Multiplicity of actions and diverse outcome. Cerebrovascular and Brain Metabolism Reviews. 1994;**6**:299-324. DOI: PMID:7533514

[65] Machha A, Schechter AN. Dietary nitrite and nitrate: A review of potential mechanisms of cardiovascular benefits. European Journal of Nutrition. 2011;**50**:293-303. DOI: 10.1007/s00394-011-0192-5

[66] Webb AJ, Patel N, Loukogeorgakis S, Okorie M, Aboud Z, Misra S, Rashid R, Miall P, Deanfield J, Benjamin N, MacAllister R, Hobbs AJ, Ahluwalia A. Acute blood pressure lowering, vasoprotective, and antiplatelet properties of dietary nitrate via bioconversion to nitrite. Hypertension. 2008;**51**:784-790. DOI: 10.1161/HYPERTENSIONAHA.107.103523

[67] Kapil V, Milsom AB, Okorie M, Maleki-Toyserkani S, Akram F, Rehman F, Arghandawi S, Pearl V, Benjamin N, Loukogeorgakis S, Macallister R, Hobbs AJ, Webb AJ, Ahluwalia A. Inorganic nitrate supplementation lowers blood pressure in humans: Role for nitrite-derived NO. Hypertension. 2010;**56**:274-281. DOI: 10.1161/HYPERTENSIONAHA.110.153536

[68] Vanhatalo A, Bailey SJ, Blackwell JR, DiMenna FJ, Pavey TG, Wilkerson DP, Benjamin N, Winyard PG, Jones AM. Acute and chronic effects of dietary nitrate supplementation on blood pressure and the physiological responses to moderateintensity and incremental exercise. American Journal of Physiology. Regulatory, Integrative and Comparative Physiology. 2010;**299**:1121-1131. DOI: 10.1152/ajpregu.00206.2010

[69] Hobbs DA, Kaffa N, George TW, Methven L, Lovegrove JA. Blood pressure-lowering effects of beetroot juice and novel beetroot-enriched bread products in normotensive male subjects. British Journal of Nutrition. 2012;**108**:2066-2074. DOI: 10.1017/S0007114512000190

[70] Kelly J, Fulford J, Vanhatalo A, Blackwell JR, French O, Bailey SJ, Gilchrist M, Winyard PG, Jones AM. Effects of short-term dietary nitrate supplementation on blood pressure, O2 uptake kinetics, and muscle and cognitive function in older adults. American Journal of Physiology. Regulatory, Integrative and Comparative Physiology. 2013;**304**:R73-R83. DOI: 10.1152/ajpregu.00406.2012

[71] Gilchrist M, Winyard PG, Aizawa K, Anning C, Shore A, Benjamin N. Effect of dietary nitrate on blood pressure, endothelial function, and insulin sensitivity in type 2 diabetes. Free Radical Biology & Medicine. 2013;**60**:89-97. DOI: 10.1016/j.freeradbiomed.2013.01.024

[72] Coles LT, Clifton PM. Effect of beetroot juice on lowering blood pressure in free-living, disease-free adults: A randomized, placebo-controlled trial. Nutrition Journal. 2012;**11**:106. DOI: 10.1186/1475-2891-11-106

[73] Christensen PM, Nyberg M, Bangsbo J. Influence of nitrate supplementation on VO_2 kinetics and endurance of elite cyclists. Scandinavian Journal of Medicine & Science in Sports. 2013;**23**:21-31. DOI: 10.1111/sms.12005

[74] Kaimainen M, Laaksonsen, Järvenpää E, Sandell M, Huopalahti. Consumer acceptance and stability of spray dried betanin in model juices. Food Chemistry. 2015;**187**:398-406. DOI: 10.1016/j.foodchem.2015.04.064

[75] Rao MA. Rheology of Fluid, Semisolid and Solid Foods: Principles and Applications. 2nd ed. New York: Springer; 2014

[76] Vasconcellos J, Silvestre DH, Baião DS, Werneck-de-Castro JP, Alvares TS, Paschoalin VMF. A single dose of beetroot gel rich in nitrate does not improve performance but lowers blood glucose in physically active individuals. Journal of Nutrition and Metabolism. 2017;**7853034**:1-9. DOI: 10.1155/2017/7853034

[77] Clifford T, Constantinou CM, Keane KM, West DJ, Howatson G, Stevenson EJ. The plasma bioavailability of nitrate and betanin from *Beta vulgaris* rubra in humans. European Journal of Nutrition. 2016;**1-10**. DOI: 10.1007/s00394-016-1173-5

[78] Kapil V, Milsom AB, Okorie M, Maleki-Toyserkani S, Akram F, Rehman F, Arghandawi S, Pearl V, Benjamin N, Loukogeorgakis S, Macallister R, Hobbs AJ, Webb AJ, Ahluwalia A. Dietary nitrate provides sustained blood pressure lowering in hypertensive patients: A randomized, phase 2, double-blind, placebo-controlled study. Hypertension. 2015;**65**:320-327. DOI: 10.1161/HYPERTENSIONAHA.114.04675

[79] Kenjale AA, Ham KL, Stabler T, Robbins JL, Johnson JL, Vanbruggen M, Privette G, Yim E, Kraus WE, Allen JD. Dietary nitrate supplementation enhances exercise performance in peripheral arterial disease. Journal of Applied Physiology. 2011;**110**:1582-1591. DOI: 10.1152/japplphysiol.00071.2011

[80] Hobbs DA, Goulding MG, Nguyen A, Malaver T, Walker CF, George TW, Methven L, Lovegrove JA. Acute ingestion of beetroot bread increases Endothelium-Independent vasodilation and lowers diastolic blood pressure in healthy men: A randomized controlled trial. Journal of Nutrition. 2013;**143**:1399-1405. DOI: 10.3945/jn.113.175778

[81] Bond Jr. V, Curry BH, Adams RG, Asadi MS, Millis RM, Haddad GE. Effects of dietary nitrates on systemic and cerebrovascular hemodynamics. Cardiology Research and Practice. 2013;**435629**:1-9. DOI: 10.1155/2013/435629

[82] Kim JK, Moore DJ, Maurer DG, Kim-Shapiro DB, Basu S, Flanagan MP, Skulas-Ray AC, Kris-Etherton P, Proctor DN. Acute dietary nitrate supplementation does not augment submaximal forearm exercise hyperemia in healthy young men. Applied Physiology Nutrition and Metabolism Journal. 2015;**40**:122-128. DOI: 10.1139/apnm-2014-0228

[83] Jajja A, Sutyarjoko A, Lara J, Rennie K, Brandt K, Qadir O, Siervo M. Beetroot supplementation lowers daily systolic blood pressure in older, overweight subjects. Nutrition Research. 2014;**34**:868-875. DOI: 10.1016/j.nutres.2014.09.007

[84] Bondonno CP, Liu AH, Croft KD, Ward NC, Yang X, Considine MJ, Puddey IB, Woodman RJ, Hodgson JM. Short-term effects of nitrate-rich green leafy vegetables on blood pressure and arterial stiffness in individuals with high-normal blood pressure. Free Radical Biology and Medicine. 2014;**77**:353-362. DOI: 10.1016/j.freeradbiomed.2014.09.02

Chapter 3

Food Preservatives from Plants

Hubert Antolak and Dorota Kregiel

Additional information is available at the end of the chapter

http://dx.doi.org/10.5772/intechopen.70090

Abstract

It has long been shown that phytochemicals protect plants against viruses, bacteria, fungi and herbivores, but only relatively recently we have learnt that they are also critical in protecting humans against diseases. A significant amount of medicinal plants is consumed by humans. As food-related products, they additionally improve human health and general well-being. This chapter deals with plant-derived food preservatives. Particular attention has been paid to the following berry fruits: cranberry (*Vaccinium macrocarpon*), bilberry (*Vaccinium myrtillus*), black currant (*Ribes nigrum*), elderberry (*Sambucus nigra*), cornelian cherry (*Cornus mas*) and açaí (*Euterpe oleracea*), as well as the following herbs and spices: peppermint (*Mentha piperita*), basil (*Ocimum basilicum*), rosemary (*Rosmarinus officinalis*), thyme (*Thymus vulgaris*), nettle (*Urtica dioica*), cinnamon (*Cinnamomum zeylanicum*) bark, cloves (*Syzygium aromaticum*) and licorice (*Glycyrrhiza glabra*) as alternative sources of natural antimicrobial and antibiofilm agents with potential use in food industry. Moreover, we present an overview of the most recent information on the positive effect of bioactive compounds of these plants on human health. This chapter is a collection of essential and valuable information for food producers willing to use plant-derived bioactive substances for ensuring the microbiological safety of products.

Keywords: plant extracts, medicinal plants, antimicrobials, antiadhesives, food preservatives, food additives

1. Introduction

According to the 'Plant List'—the first consolidated checklist of the world's plants completed in 2010—there are up to 1 million plant species on Earth, of which around 350,000 have accepted names. Due to the fact that new species are still being identified, calculating anything like an accurate number is further complicated by many examples of the same species in different areas being known by different names. It is estimated that the total number of plants is of the order of 400,000 species. Despite such a great biodiversity, only 80,000 are edible for humans and animals, of which 30 produce 95% of human calories [1, 2]. Furthermore, a significant number of known plants are not only a source of nutrients but also find use as remedies for health problems. For centuries, plants have been known as a sources of bioactive compounds usable to fight health issues. According to the World Health Organization report released in 2003, over 50% of the population of Europe, North America and other industrialized regions have used complementary or alternative medicine at least once. What is more, traditional herbal preparations account for 30–50% of the total medicinal consumption in China. Moreover, the global market of herbal medicines stands at over $ 60 billion annually and generates increasing interest [3].

2. Medicinal plants through ages

The importance of medicinal plants was acknowledged at least 50,000 years ago, as evidenced by numerous archaeological excavations. However, the oldest known medical document—the Ebers papyrus—is dated to the fifteenth century BC. This document contains more than 800 recipes of various medicines from herbs: extracts, lotions and liniments [4, 5]. It is assumed that the oldest document describing medicinal plants used in China and the Far East is the Pent-Sao book. A copy of this chapter, dated seventh century AD, describes nearly 400 herbs from which juices, infusions and ointments were produced. It is worth mentioning that one of the most respected plant materials was Chinese ginseng (*Panax ginseng*), considered to be a drug for immunity, strengthening and energizing [6]. India is also known for its traditional medical systems. One of them—Ayurveda—is found mentioned in the ancient Vedas. The Ayurvedic concept developed between 2500 and 500 BC. The name means 'science of life', and Ayurvedic is also called the 'science of longevity'. The concept is based on the natural treatment methods to cure many common diseases such as food allergies. In general, Ayurvedic is a system based on health care and long life [7]. Herbal medicine in Europe was discovered much later, and the cradle of this field was Greece. Hippocrates of Kos is considered to be the father of phytotherapy, as his work *Corpus Hippocratium,* released after his death, contained information about the beneficial effects of more than 400 plants [8]. In later years (370–287 BC), Theophrastus of Eresos, considered to be the father of botany, described more than 500 plants. However, much more important, in this field, was a five-volume work of Dioscorides—*De Materia Medica* (40–90 AD), considered as one of the most prominent books on herb treatment [9]. On the other hand, the most famous of the Romans was Claudius Galenus, called Galen (130–200 AD), who described 450 plants, claiming that the health effect depends on the form in which the medication is taken. Until the fifteenth century, medicine was based on the recipes described by Galen, and today, medications obtained by crushing and extraction of the plant material are called 'galenic' [10]. However, with the fall of the Roman Empire, the development of herbal medicine has been slowed down. In later centuries, the Arabs introduced new medicinal plants and forms of medicine such as spirit-based syrups. They were the first one to begin using natural dyes and flavourings such as cloves, vanilla, camphor and nutmeg. Avicenna was considered to be the father of medicine of those times—the author of nearly

500 books, among them the *Canon of Medicine* which includes more than 700 herbal medicines [11]. Five centuries later, Paracelsus (1493–1541 AD) disagreed with the views of Hippocrates, Galen and Avicenna. In his opinion, life processes have the character of a chemical transformation; thus, the whole plant is not needed in treatment of a disease, only the specific substance. In that way, phytotherapy was divided in two directions: Galleons—based on extracts, ointments, juices and stocks—and Paracelsus—based on chemical compounds extracted from plants. As a result, a few centuries later, in 1804, the first alkaloid—morphine—was extracted. To this day, both directions are extremely important elements in medicine and daily life [12, 13]. As a consequence, scientists worked not only on the separation of chemical compounds from plants but also on their chemical synthesis. For example, despite the fact that salicylic acid was obtained in a form of an extract from a bark of a willow, the component was synthesized in 1859 by Hermann Kolbe. His discovery is considered to be the beginning of the pharmaceutical industry [14].

The field of pharmaceutical drugs as we know today is based on the historical use of plants. Valuable sources of that are ancient and medieval texts written by explorers who had a combined interest in botany and medicine. The ethnopharmacological knowledge gained from indigenous peoples of a particular region is still used in the search for new medicines. Compounds originating from the plant kingdom of the world make up a framework from which novel drugs are developed. What is more, the importance of plants and the biodiversity also results from the fact that bioactive compounds found in the wild may not be reproducible in the laboratory [15]. The example of this strategy is the identification of artemisinin–antimalarian agent for the discovery of which Tu Youyou received the Nobel Prize in Physiology or Medicine in 2015 [16]. Sweet wormtree (*Artemisia annua*)—the plant, from which the compound was extracted–has been well known in the Chinese medicine since 200 BC. A great number of traditional drugs commonly used in Western medicine are derived from plants, so it comes as no surprise that plants remain an important source of starting material for discovery and commercial use. What is more, plants have attracted scientific interest because 60% of the antimicrobial drugs discovered in the past few decades are of natural origin [17].

It has long been shown that phytochemicals protect plants against viruses, bacteria, fungi and herbivores, but only relatively recently, we have learnt that they are also critical in protecting humans against diseases. Significant part of medicinal plants is consumed by humans, and as a food, it additionally improves human health and well-being in general. It is well known that a diverse array of herbs, vegetables, fruits and grains, besides having nutrients, vitamins and minerals, also possess a large variety of biologically active compounds. These bioactive components as well as their sources as a functional food have recently gained much attention and publicity. The term 'functional food' was first introduced in Japan and refers to foods which, in addition to basic essential nutrients, also contain ingredients beneficial to human health that for example reduce the risk of chronic diseases, promote health and extend longevity [15]. Therefore, the primary function of food is not only to satisfy hunger and to provide the necessary nutrients but also to prevent diseases and to improve physical and mental well-being. What is more, a growing number of consumers are becoming aware of functional foods and its beneficial properties. Therefore, it is considered that functional food is a long-term trend with an important market potential, which is conditioned by the expectations of consumers. Taking this into account, more and more recent innovations are increasingly being used. Food industry innovations can be classified as the following: (1) new food ingredients and materials, (2) innovations in fresh foods, (3) new food processing techniques, (4) innovations in food quality and (5) new packaging methods [18]. As a result, more and more completely new products based on natural additives, in the form of fruit juice or herbal extracts, previously considered as medicinal or unattractive in terms of sensory, are available on the market. Moreover, food manufacturers have realized that besides the fulfilment of health-related consumer expectations, they can use the natural source of bioactive compounds for their purposes [19]. With the increased negative attitudes of consumers to chemical food additives such as preservatives, acidity stabilizers or food colourings, fruit juices and herbs extracts have gained in importance fulfilling all of these functions. The application of natural plant food preservatives with additional potential as health-promoting agents is especially interesting [20]. What is more, natural plant-origin, antimicrobial compounds have been investigated as alternatives to synthetic ones for preserving food quality, owing to their effectiveness against food spoilage and foodborne pathogens [21].

3. Bioactive compounds

Metabolites produced by each living cell can be generally divided into two groups: primary metabolites (PMs) and secondary metabolites (SMs). PMs are the chemicals aimed at growth and development and include carbohydrates, amino acids, proteins and lipids. SMs are characterized as compounds believed to help plant to increase overall ability to survive and overcome local challenges by allowing them to interact with their surroundings. Most of bioactive compounds of plants are produced as secondary metabolites, giving plants their colour, flavour and aroma. The simplest definition of plant origin bioactive compound is 'secondary plant metabolites eliciting pharmacological or toxicological effects in human and animals'. Bioactive compounds are present in all plant material: vegetables, grains, legumes, beans, fruits, herbs, roots, leaves and seeds. They are largely responsible for the medicinal properties and health benefits of herbs, but also for poisonous and toxic effects of others [22, 23].

Despite the fact that classification of bioactive compounds in different categories and subcategories is still inconsistent, they can be divided into three main categories: (1) terpenes and terpenoids (25,000 types), (2) alkaloids (12,000 types) and (3) phenolic compounds (PC) (8000 types) [24]. Basing on unique structural characteristics, and thus on the way of their biosynthesis, as well as function, bioactive compounds belong to one of a number of families. There are four major pathways for biosynthesis of SM: (1) shikimic acid pathway; (2) malonic acid pathway; (3) mevalonic acid pathway and (4) non-mevalonate pathway [25]. Alkaloids are generally produced by aromatic amino acids (shikimic acid pathway) and by aliphatic amino acids. PC are synthesized through two pathways: shikimic acid pathway and malonic acid pathway, while terpenes are synthetized through mevalonic and non-mevalonate acid pathways [23]. The overall classification of bioactive compounds of plants with examples is shown in **Figure 1**.

http://dx.doi.org/10.5772/intechopen.70090

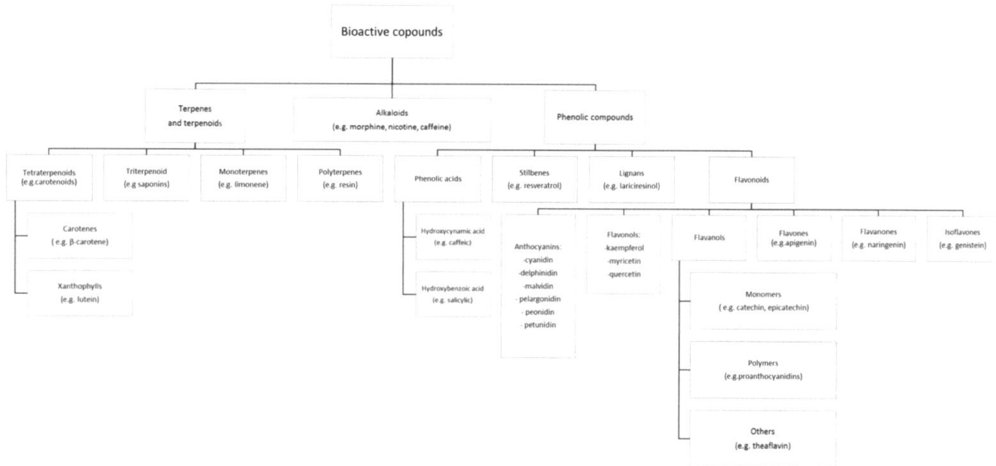

Figure 1. General classification of bioactive compounds of plants.

Phenolic compounds are widely distributed, and an important group of compounds occurs in plants. Polyphenol family contains about 8000 structurally different compounds, commonly found in fruits, vegetables, seeds, flowers and leaves. They are generally categorized as phenolic acids and derivatives, flavonoids, tannins, stilbenes, lignans, quinones and others based on the number of phenolic rings and of the structural elements that link these rings. Biosynthesis of mono and polyphenolic compounds is carried out from carbohydrates by way of shikimic acid, phenylpropanoid and flavonoid pathways [26]. Generally phenolic acids contain two distinguishing constitutive carbon frameworks: the hydroxycinnamic and hydroxybenzoic structures. The first group of phenolic acids includes ferulic, chlorogenic, sinapic, caffeic and p-coumaric acids, while the second one contains gallic, syringic, protocatechuic and vanillic acids. Other polyphenols are also considered as phenolic acids: capsaicin, rosmarinic acid, gingerol, gossypol, ellagic acid and cynarin [27]. It is worth noting that caffeic, p-coumaric, vanillic, ferulic and protocatechuic acids are widely distributed in nearly all plants. Red fruits, especially berries: blueberry (*Vaccinium corymbosum*), blackberry (*V. myrtillus*), chokeberry (*Aronia melanocarpa*), strawberry (*Fragaria virginiana*), red raspberry (*Rubus idaeus*), elderberry (*S. nigra*) and black currant (*R. nigrum*), are a rich source of phenolic acids (**Figure 2**).

Another group of phenolic compounds are **flavonoids** which include flavonols, flavanols, flavones, flavanones, isoflavones and anthocyanins [28, 29]. The most widespread of all the groups are flavonols conjugates with over 200 different sugar conjugates of kaempferol. In

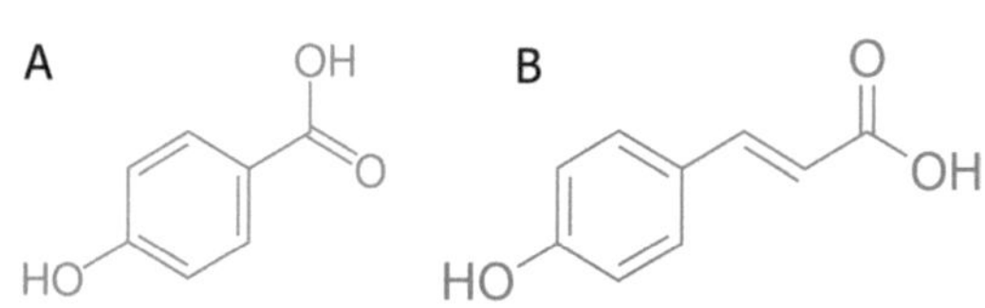

Figure 2. Representatives of phenolic acids. A—4-hydroxybenzoic acid; B—coumaric acid.

addition to kaempferol, flavonols also include myricetin, quercetin, morin, galangin and isorhamnetin and most commonly occur in the form of O-glycosides (**Figure 3**). They are present in a wide range of food such as fruit (cherries, blueberries, apples), vegetables (broccoli, tomato), beverages (red wine), herbs and spices (caraway, cumin).

On the other hand, flavones appear in a limited number of raw plant materials and can be found in parsley, celery, thyme, tea, legumes and certain other herbs. The major flavones are apigenin, luteolin, baicalein, chrysin and their derivatives. Tangeretin, nobiletin and sinensetin, the most hydrophobic of all of the flavonoids, also belong to this group [26]. Naringenin and hesperetin as well as their glycosides (naringin, hesperidin) are flavanones present mainly in citrus fruits (oranges, lemons), grapes, and medicinal herbs belonging to the family: *Rutaceae, Rosaceae*, and *Leguminosae* [30]. Another, very important group are flavanols which occur as simple monomers of (+)-catechin or (-)-epicatechin, as well as in hydroxylated (gallocatechins) and esterified (gallic acid) forms (**Figure 4**).

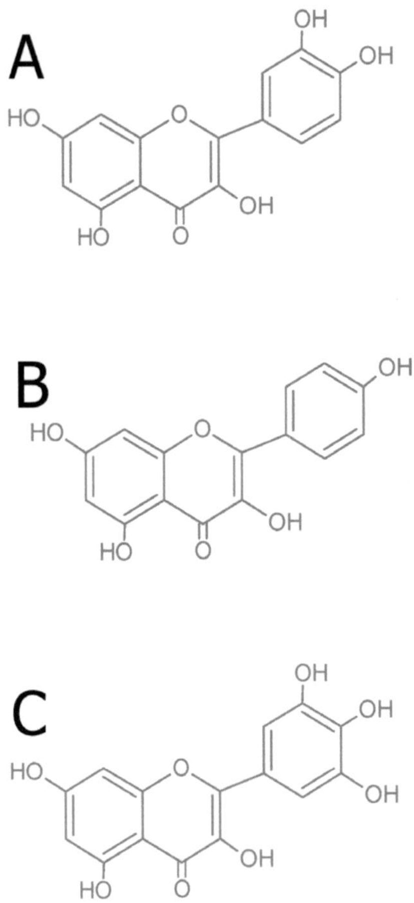

Figure 3. Typical flavonols: A—quercetin; B—kaempferol; C—myricetin.

http://dx.doi.org/10.5772/intechopen.70090

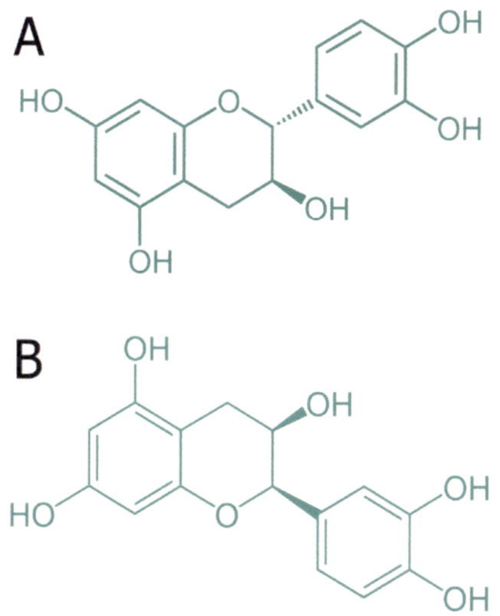

Figure 4. Structures of A—(+)- catechin; B—(-)-epicatechin.

The most important among flavanols are catechin, epicatechin, epigallocatechin, epicatechin gallate and epigallocatechin gallate. They are present in tea (mainly green tea), apples, grapes, berries and cocoa. Anthocyanins, including cyanidin, delphinidin, malvidin, peonidin and pelargonidin as well as their glycosides, are widely distributed in fruits giving them a characteristic colour depended on their pH. Therefore, in addition to health properties, they arouse interest as food colourings [31] (**Figure 5**).

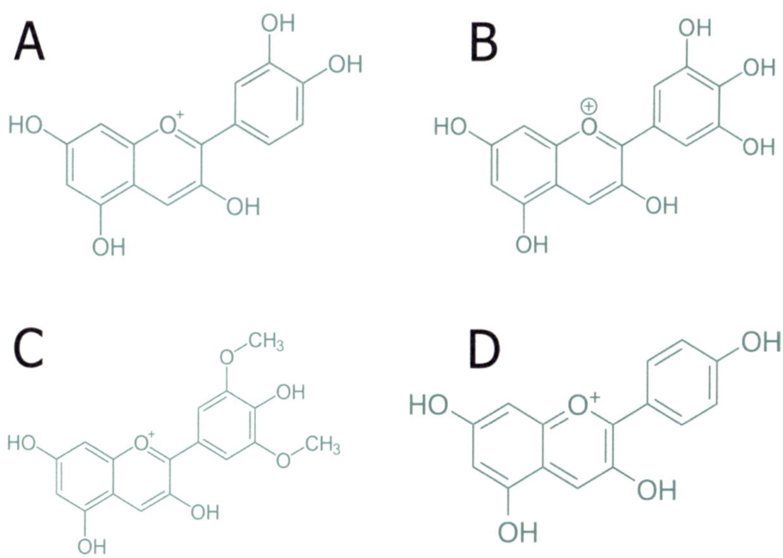

Figure 5. Representatives of anthocyanins: A—cyanidin; B—delphinidin; C—malvidin; D—pelargonidin.

Alkaloids are the next class commonly found in plants. General definition has been suggested in 1983 by Pelletier, and it was 'cyclic compound containing nitrogen in negative oxidation state which is of limited distribution in organisms'. Despite the diversity of alkaloids, they show a similarity in the chemical structure: (1) alkaloids contain nitrogen, in most cases derived from several amino acids; (2) exhibit alkaline pH and (3) they have no basic forms like quaternary compounds and N-oxides [32]. The most often used classification of alkaloids is so-called (1) true alkaloids—atropine, nicotine and morphine; (2) protoalkaloids—adrenaline and ephedrine; and (3) pseudoalkaloids—caffeine, theobromine and theacrine. On the other hand, based on the biosynthesis, alkaloids can be divided into indole alkaloids (tryptophan derived, e.g. ergometrine), piperidine (lysine derived, e.g. lobeline), pyrrolidine (ornithine derived, e.g. hygrine), phenylethylamine (tyrosine derived) and imidazole (derived from histidine, e.g. pilocarpine). It is believed that about 14–20% of plant species contain alkaloids. Main representatives are plants belonging to the families: *Solanaceae* (e.g. *Nicotiana tabacum* and *Datura stramonium*), *Papaveraceae* (e.g. *Papaver somniferum*), *Ranunculaceae* (e.g. *Hydrastis canadensis*), *Erythroxylaceae* (e.g. *Erythroxylum coca*), *Rubiaceae* (e.g. *Pausinystalia johimbe*) and *Campanulaceae* (e.g. *Lobelia inflate*). Alkaloids in plants are considered to be a growth regulation factor, serve as a reserve substance as well as play an important role in plants [33]. Alkaloids are generally used in therapeutics and pharmacology. They show a wide range of biological properties: anti-inflammatory, antidepressant, antitumor, antiviral, antihypertensive, but also antimicrobial and antimalarial [34].

Terpenes, also called terpenoids, are the most diverse class of natural bioactive compounds. It is believed that this class can count up to 40,000 different chemicals [35]. They can be classified into many categories based on the number of carbon atoms as well as the presence of isoprene residues (IPR): (1) monoterpenes consist of 10 carbon atoms, or two IPR; (2) sesquiterpenes contain 15 carbon atoms, or three IPR; in the composition of (3) diterpenes consist of 20 C atoms, or four IPR; (4) triterpenes consist of 30 C atoms, or six isoprene units, while in the composition of (5) tetraterpenes there are 10 more carbon atoms, or two more isoprenes. The last group polyterpenes are composed of large number of IPS. Although terpenes are secondary metabolites, they have a well-characterized function in plants growth. For example, gibberellins (diterpenes) are plant hormones, sterols (triterpene derivatives) are responsible for cell stabilization, while carotenoids show protective activities against photo-oxidation. The best known among this class of compounds are carotenes and oxycarotenoids (xanthophylls) belonging to tetraterpene family. Carotenoids, such as β-carotene, lutein, lycopene or zeaxanthin, are lipid-soluble colour pigments occurring in vegetables and fruits, giving them a yellow, orange and even a red colour [26]. On the other hand, some terpenes are toxins and play important defensive roles against many insects and mammals. Pyrethroids and monoterpene esters, from *Chrysanthemum* spp., show strong insecticidal activities. However, it is believed that the most important from all of the terpenes are volatile monoterpenes and sesquiterpenes known as essential oils (EOs). They occur mainly in herbs and spices, as well as some fruit, giving them a characteristic aroma. Peppermint (*M. piperita*), lemon (*Citrus limon*), basil (*O. basilicum*), cinnamon (*C. zeylanicum*) and rosemary (*R. officinalis*) are the examples of plants that are rich sources of essential oils. EOs are well-known bioactive compounds used in aromatherapy, microbiology and agroindustry. It has been shown that terpenes exhibit various pharmacological properties such as anti-inflammatory, anticarcinogenic, antitumor, antibacterial, antimalarial, antiviral, antibacterial as well as hepatoprotective [24–26].

Saponins are a group of compounds which attract attention. Structurally, they can be divided into triterpenoids or steroids. They contain a triterpene or steroid aglycone and attached sugar chain(s), mainly consisting of rhamnose, xylose, arabinose, galactose and fucose. Their amphoteric nature results from the presence of hydrophobic aglycone and hydrophilic sugar chains [36]. These molecules are responsible for the unique properties of saponins: emulsifying, foaming and detergenic properties. The source of steroidal saponin in nature is *Lilianae* (monocotyledons), while triterpenoid is commonly found in *Dicotyledoneae* (dicotyledons). In plants, they accumulate mainly in the bark, roots and leaves of the plants belonging to *Agavaceae, Leguminosae, Rosaceae, Caryophyllaceae* and *Umbelliferae* families [37]. The main representatives of plants characterized by a rich composition of saponins are as follows: Chinese honey locust (*Gleditsia sinensis*), soapwort (*Saponaria officinalis*), Mongolian milkvetch (*Astragalus propinquus*), ginger (*P. ginseng*), Yucca (*Yucca schidigera*) and soapbark (*Quillaja saponaria*). The last two are traditionally used as detergents (*Q. saponaria*) or in animal nutrition in order to reduce faecal odours (*Y. schidigera*). Saponins are compounds exhibiting insecticidal, anthelmintic, molluscicidal as well as antiviral, antibacterial and antifungal activities in plants [38]. Saponins show anticancer properties, while these from *Androsace umbellate* are reported as inductors of cell apoptosis and inhibitors of cancer cells [39, 40]. What is more, saponins from *Yucca* show an inhibition of food-spoilage yeast *Debaryomyces hansenii, Pichia nakazawae, Zygosaccharomyces rouxii, Candida famata* and *Hansenula anomala* [41]. In recent years, saponins sparked interest as a natural, environment friendly additives/ingredients. For example, *Y. schidigera* are commonly used in Japan as an additive against yeasts contaminating cooked rice, pickled vegetables or fish meat [42]. Moreover, it was found that yeast treated with saponin extract from *Q. saponaria* showed increased cell membrane permeability, as a result facilitating the preparation of yeast salt-free lysates much easier [43].

Due to the richness and variety of bioactive substances contained in plants and their positive effects on human health, they constitute important raw materials used in a variety of industries. In view of their promoting properties, such as antioxidant, anti-inflammatory, anticancer, antimutagenic, antiallergenic, antifungal, antibacterial and many others, research has been intensified towards fruit, vegetables, plants, herbs and spices which will provide an attractive and functional addition to food products.

4. Herbs and spices

Herbs and spices have been playing a key role in the daily life of mankind since the ancient times. They are derived from various parts of plants from all over the world. Herbs are generally fresh or dried leaves of plants, while spices are divided into four groups: (1) pungent species, (2) aromatic seeds, (3) aromatic barks and (4) coloured species [44]. It is considered that about 400 herbs and spices are commonly used around the world. It is speculated that a much larger proportion of human population uses herbs and spices as therapeutic remedies, compared to prescription pharmaceuticals. The characteristics of herbs are conditioned by their chemical composition, e.g. its aroma and flavour depend on the volatile oils such as monoterpenes [45, 46]. Among other things, these features made them recognized as a food, food additives and in the manufacturing of cosmetics and pharmaceuticals. What is more, global awareness of health and environmental issues, especially in the developed countries, caused increased demands for medical herbs and spices as well as food products containing these plant materials. Increasingly common are new herbs or mixtures of herbs designed to improve the health of consumers. Due to the popularity of functional and ecological foods, and as well as increasing consumer awareness about the impact of diet on human health, the pro-healthy herbs are used in food production [46].

There is no doubt herbs and spices of natural origin show the extraordinary properties and advantages for human population. Due to their natural origin, they are generally considered as safe for consumption and thus can be used in food production. The consequence of rising consumer awareness is the demands for high-quality foods. What is more, there are also concerns about microbiological food safety due to the occurrence of new food-borne pathogens and spoilage microorganisms. Therefore, food manufacturers are aware of the limited use of synthetic preservatives, and they are looking for new sources of bioactive compounds that exhibit three roles: (1) meet the expectations of the new trends of healthy food; (2) will be characterized by antimicrobiological properties and (3) show other functions of food additives (acidity stabilizers, dyes or other) [47]. Further part of the chapter provides an overview of the literature on the properties of selected herbs and spices, which are a potential source of food additives. Herbs have been chosen with regard to their antimicrobial properties and health benefits of their regular consumption.

4.1. Peppermint (*M. piperita*)

Description. Peppermint (*M. piperita*) is a natural hybrid of watermint (*Mentha aquatic*) and spearmint (*Mentha spicata*) belonging to family *Lamiaceae*. The herb is native to Europe, but nowadays, it is commonly found around the world and used as a pharmaceutic and cosmetic material, as well as a food additive, mainly in order to enrich the products in fresh, mint flavour and aroma. In traditional medicine, peppermint essential oils are commonly used. It is commonly found as one of the compounds of tea, but the extract of peppermint is also consumed [48]. *M. piperita* leaves are popularly known as refreshing additive in beverages. According to Food and Agricultural Organization of United Nations (FAO), world's production of peppermint in 2010 reached more than 80,000 tons. The biggest global producer was Africa—89% of the amount [49].

Bioactive compounds *M. piperita* is mostly known for its essential oils, and, hence, volatile components, which are often used in cosmetics, pharmaceutics and food industry. It is considered that major compounds of peppermint EOs are menthone (approximately 30%) and menthol (25%). Other compounds like menthofuran, limonene, β-phellandrene, isomenthone, menthol acetate, pulegone, β-caryophyllene, neomenthol and germacrene D are encountered at much lower concentration [50, 51]. Phytochemicals detected in the *M. piperita* extracts are also rosmarinic acid, ferulic acid, gallic acid, vanillic acid, p-coumaric acid, caffeic acid, syringic acid, (+)-catechin, (-)-epigallocatechin gallate, eriocitrin, hesperidin and luteolin-7-O-rutinoside [48, 52]. Generally, 50% of all of the *M. piperita* bioactive compounds are flavonoids, followed by about 42% phenolic acids, and 2.5% of lignans and stilbenes. According to Fecka and Turek, 2 g of peppermint can provide 88 mg of phenolic acids, but the value depends on the type of the product [53].

Health benefits Research on the health-benefiting activities of peppermint showed that one in eight plants exhibits strong activity in suppressing the effect of okadaic acid which promotes tumour formation [54]. Extract of *M. piperita* suppresses mutagenicity of human carcinogens formed in cooked meat. What is more, methanol extract from peppermint is cytotoxic to L1210 cancer cells. Extract of mint also reduces lungs carcinogenicity and mutagenicity [55]. Other studies suggest that *M. piperita* can affect the bioavailability of certain drugs. Furthermore, flavonoids from this herb show antiallergic effects in research with rat peritoneal mast cells. It is believed that luteolin-7-O-rutinoside is an especially potent compound. Both the aqueous extracts and peppermint oils exhibit potent antiviral properties towards herpes simplex virus (HSV), influenza, vaccinia virus, suppressing replicative ability of HSV-1 [56]. It has been found that virulence of both herpes simplex virus 1 and 2 is inhibited by peppermint oil. It has also been shown that bioactive compounds contained in herbs are characterized by gastrointestinal activities (stimulating choleretic activity), antiallergenic actions inhibiting sneezing and with menthol from peppermint a significant enhancement of the nasal sensation [52].

Antimicrobial activity It has been documented that *M. piperita* extracts and EOs are characterized by broad spectrum of antibacterial activities against gram-positive and gram-negative pathogens as well as antifungal activities against yeasts and moulds. In the study of Singh et al., it was found that gram-positive bacteria such as *Staphylococcus aureus* and *Staphylococcus pyogenes* are more sensitive to essential oil compared to *Escherichia coli*. Authors also established that the growth inhibition was compared with gentamycin [57]. Mint oil also had bactericidal effect against *Staphylococcus mutans, Salmonella typhimurium, Pseudomonas aeruginosa* and *Shigella* spp. It has been confirmed that gram-negative bacteria such *P. aeruginosa* are less sensitive to mint oil that other tested bacterial strains [58]. On the other hand, Pramila et al. noted that methanol extract of *M. piperita* showed stronger activity against *E. coli* compared to *Staphylococcus* and *Acinetobacter* strains [59]. Mint oil was also effective in inhibiting *Salmonella enteritidis* in cucumber salads, tzatziki yoghurts, as well as cod's roe salad [60]. The *M. piperita* EOs and ethanol extract showed antifungal activity against *Candida* spp.: *Candida tropicalis, Candida albicans, Candida glabrate* and *Candida parapsilosis*, but its infusion did not have any antifungal properties not only to them but also moulds: *Aspergillus niger, Aspergillus flavus, Aspergillus parasiticus, Rhizopus solani, Alternaria alternata* [61]. What is more, peppermint oils and extract showed anti-adhesive and antibiofilm properties against food spoilage bacteria *Asaia lannensis* and *Asaia bogorensis* as well as rods *P. aeruginosa*, bacilli *Listeria monocytogenes* and yeasts *C. albicans, Candida dubliniensis* [62–65].

4.2. Basil (*O. basilicum*)

Description Another representative of a medical herb belonging to the family *Lamiaceae* is basil (*O. basilicum*), originating from the warm tropical climates of India, Africa, Asia as well as Mediterranean Europe. *O. basilicum* is one of the 150 species, widely cultivated worldwide, due to its use in cooking and folk medicine for treating headaches, coughs and kidney malfunctions. In everyday life, basil is a popular flavouring agent, an additive to medications, cosmetics, perfume and food products [66, 67].

Bioactive compounds Similar to peppermint (*M. piperita*), basil belongs to the aromatic plants, and therefore, the most common mixture of bioactive compounds derived from this plant is essential oils. EOs of *O. basilicum* mainly contain eugenol, estragol, methyl cinnamate, linalool, geranial and neral, but the composition can differ dependent upon harvest dates and growth condition. Other compounds identified in basil oils are E-caryophyllene, aromadendrene, α-humulene, terpinen-4-ol, γ-terpinene and camphor [66]. Typical methanolic basil extracts mainly contain: rosmarinic acid—as the dominant phenolic acid, chicoric acid—recently discovered in basil leaves, and caffeic and caftaric acids—at lower concentrations [67, 68]. On the other hand, ethanolic extracts of basil, in addition to rosmarinic, chicoric and caftaric acids, contain chlorogenic, gentisic, ferulic and p-coumaric acids, as well as β-carotene and lutein-zeaxanthin. In the study of Vlase et al., it was documented that the extract is a source of quercetin-3-glucoside, quercetin-3-rutinoside, quercetin-3-rhamnoside as well as luteolin [69].

Health benefits Due to the phenolic compounds as well as essential oils, basil is widely used as an anti-inflammatory, insecticidal and nematicidal agent. Traditionally, decoction of *O. basilicum* root is used as a drink for stomach pains and as an enema in constipation. On the other hand, tea from basil leaves is commonly inhaled for nasal and bronchial catarrh, and as a sudorific and stomachic agent [66]. What is more, some scientific literature reports that basil shows antidiabetic, adaptogenic, cardioprotective, immunostimulatory, anticarcinogenic and hepatoprotective properties [70].

Antimicrobial activity The ethanolic extracts of *O. basilicum* showed stronger antibacterial activity against gram-positive *S. aureus* than against gram-negative *L. monocytogenes, E. coli, S. typhimurium* and yeasts *C. albicans* [71]. On the other hand, the results obtained for essential oils suggest that gram-negative bacteria such *E. coli* and *P. aeruginosa* are characterized by higher sensitivity to bioactive compounds from basil [71]. Sienkiewicz et al. documented that basil oil exhibited strong antibacterial activity against clinical strains of *E. coli*, including β-lactamase positive [72]. It is considered that mainly linalool and eugenol are responsible for the antibacterial activity of basil EOs. What is more, essential oils from basil show antifungal activity against common plant pathogens: *Glomerella cingulata, Fusarium solani, A. alternata* and *Fulvia fulva* [73]. Basil EOs were tested as a natural fungicide against *Penicillium* and *Aspergillus* strains isolated from sausage [74]. Antimicrobial activity of *O. basilicum* essential oils was also shown against *S. enteritidis* both *in vitro* and in a food model by Rattanachaikunsopon and Phumkhachorn [75]. The results obtained by Carovic-Stanko et al. indicated that *Ocimum americanum* shows antimicrobial activity, against *E. coli* 0157:H7, *Enterobacter faecalis, Enterobacter faecium, Proteus vulgaris, S. aureus* and *Staphylococcus epidermis*

[76]. Due to the wide use against bacteria and fungi, basil oil is widely considered as a natural agent improving food safety.

4.3. Rosemary (*R. officinalis*)

Description Rosemary (*R. officinalis*), 1 m high, evergreen shrub with upright stems, whitish-blue flowers and dark green leaves, belongs to the family *Lamiaceae*. The herb is commonly cultivated in Mediterranean countries: Spain, Tunisia, France and Italy. Due to its intense pleasant smell reminiscent of pine, rosemary is widely used in traditional medicine and cosmetics and as a food flavouring [77].

Bioactive compounds The chemical analysis of the essential oil obtained from rosemary indicated that the herb is a rich source of monoterpenes and contains mainly: eucalyptol (1,8-cineole), camphor, β-pinene, borneol, limonene as well as camphene. What is more, according to Kontogianni et al., *R. officinalis* also contains isorhamnetin-3-O-hexoside, rosmarinic acid, carnosic acid and the triterpenic acids (oleanolic, ursolic, betulinic), as well as homoplantaginin [78]. On the other hand, extract of rosemary contains flavonol (isorhamnetin), flavones (luteolin, apigenin, hispidulin) and phenolic acids (hydroxybenzoic, hydroxycinnamic derivatives). Additionally, in the chapter of Mena et al., luteolin-rutinoside, luteolin-hexoside, isorhamnetin-3-O-hexoside, isorhamnetin-rutinoside, eriodictyol, luteolin hesperetin and epirosmanol have also been detected [79].

Health benefits Rosemary is generally an important source of bioactive compounds exhibiting several health-benefiting activities. In traditional medicine, oils, water and ethanolic extract of rosemary are commonly used as an agent for digestive, astringent, diuretic and diaphoretic problems, as well as mild analgesic and for physical and mental fatigue. Moreover, phytochemical composition contained in the herb showed anti-inflammatory, antidepressant, antiatherogenic, hepatoprotective, nephroprotective, antiobesity and anticancer properties. It has been documented that bioactive compounds, mainly carnosic acid, rosmarinic acid and carnosol, showed high antioxidant activity. What is more, essential oil of rosemary, due to monoterpenes such as 1,8-cineole, camphor and pinene content, is used as an antimicrobial biopreservative in food production [79–82].

Antimicrobial activity High bioactive compound content in essential oils and extract has been documented as a strong antimicrobial agent. Rosemary EOs exhibit antifungal activity against *Fusarium graminearum* (cereal pathogen), α-pinene and 1,8-cineole contained in the oil show activity against, *C. albicans*, and *A. niger* [81]. Rosemary oil also exhibits bacteriodical and bacteriostatic activity against *S. aureus*, *Bacillus subtilis* and *P. aeruginosa* [83]. The antibacterial, fungistatic and fungicidal activities of *R. officinalis* were also noted in the article of Bozin et al. The authors showed that rosemary EOs had strong activities against *C. albicans*, *Trichophyton tonsurans*, *Trichophyton rubrum*, *Epidermophyton floccosum*, *Microsporum canis*, *P. aeruginosa*, *E. coli*, *Salmonella typhi*, *S. enteritidis*, *Shigella sonnei*, *Micrococcus flavus*, *Sarcina lutea*, *S. aureus*, *Staphylococcus epidermidis* and *B. subtilis* [84]. Antifungal activity of rosemary was also studied against fresh dough spoilage *Aspergillus* spp. and *Penicillium* spp., and microencapsulation retained the antimicrobial property of the EOs [85]. Antimicrobial activity of essential oils depends on their volatile chemicals, such as aromatic compounds and polyphenols. The mechanism of action includes direct activity on cytoplasmic membrane, causing changes in lipid bilayer, and as a result disrupting membranes, which finally leads to leakage of the cell content [58].

4.4. Thyme (*T. vulgaris*)

Description Thyme (*T. vulgaris*), also called as common thyme, German thyme or garden thyme, is the next representative of the *Lamiaceae* family. This herb is native to Mediterranean countries in Europe. Thyme is a woody-based evergreen herb with aromatic green leaves and pink flowers. *T. vulgaris* is used as a groundcover in gardens, but is much more valuable as a flavouring to rabbit, boar, and lamb meats, giving them a spicy taste. In folk medicine, thyme was used as an infusion or an additive for baths for treatment of skin diseases, as well as carminative, and a sedative medicament [86].

Bioactive compounds Similar to other aromatic herbs, bioactive compounds from thyme can be obtained, in both aqueous form or alcoholic extracts and essential oils. *Thymus* EOs are generally a mixture of monoterpenes, mainly thymol (approximately 50% of all of the compounds) and carvacrol (10%). Essential oil of thyme is also a source of α-terpineol, γ-terpinene, linalool, camphor and caryophyllene [86]. What is more, phenolic acids (caffeic, p-coumaric, cinnamic, carnosic, rosmarinic, caffeoylquinic, ferulic and quinic), as well as flavonols (quercetin-7-O-glucoside), flavones (apigenin), flavanones (naringenin) can also be present in methanolic extracts [87]. Steroids, saponins, alkaloids, tannins and flavonoids can also be extracted from thyme using hexane, ethyl acetate and butanol [88].

Health benefits Infusions and decoctions of thyme leaves have been used for thousands of years for treatment of a cold, in a production of tonics, as a medicine in digestive problems, as antispasmodic, expectorant in upper respiratory tract infections, as well as an carminative agent. The herb is also reported to enhance the activity of the superoxide dismutase, an enzyme which has the potential to act as an anti-inflammatory agent [89]. What is more, scientific reports showed that *T. vulgaris* possesses numerous biological properties including antioxidant, antimicrobial and sedative and can be used in gastroenteric and bronchopulmonary disorders [90]. What is more, thymol EOs show antiseptic properties 30 times higher than phenol and thus are used as the main active antiseptic ingredient in chemotherapeutic mouth rinses against gingivitis.

Antimicrobial activity Investigating the antimicrobial activity of thyme essential oil showed that bioactive components are a strong antibacterial and antifungal agents against both food spoilage microflora and photogenic microflora. Essential oil of thyme prevented the growth of gram-negative bacteria *Erwinia amylovora* which is responsible for fire blight disease of apples and pears [91]. Thyme and EOs demonstrated strong inhibitory effects against *Colletotrichum gloeosporioides* and *Rhizopus stolonifera* responsible for the spoilage of storage papaya fruits [92]. In the chapter of Arras and Usai, essential oil of *Thymus* sp. showed strong fungitoxic activity against citrus pathogens: *Penicillium digitatum, Penicillium italicum, Botrytis cinerea* and *Alternaria citri* [93]. Satya et al. showed the inhibitory effect on *A. niger, Cryptococcus neoformans* and *C. albicans*. It is speculated that camphor contained in the oil is mainly responsible for antifungal activity of thyme oil [94]. What is more, the oil vapour of the herb-reduced peach brown rot caused by *Monilinia laxa* increasing the activity of phenylalanine ammonia lyase [95]. *T. vulgaris* extract also showed strong antibacterial activity against food pathogens such as *B. subtilis, Enterobacter cloacae, S. aureus, S. epidermidis, Salmonella typhimurium, S. enteritidis, P. aeruginosa, E. coli, M. flavus and Micrococcus mirabilis* [96].

4.5. Nettle (*U. dioica*)

Description Common nettle (*U. dioica*) is an, up to 2 m high, herbaceous perennial plant belonging to the family *Urticaceae*. The herb is native to Africa, Asia, North America and Europe, but nowadays, it is found worldwide. The name of this green plant comes from *uro*, meaning 'burn', or *urere* meaning 'to sting'. Nettle is characterized by hollow stinging hairs called trichomes occurring on leaves and stems. Trichomes act like needles which in contact with human skin inject acetylcholine, histamine, serotonin, moroidin and formic acid, causing burning and rashes. *U. dioica* has been traditionally used as a source of medicine, food and feed additive and fibres [97].

Bioactive compounds The main source of the bioactive compounds is leaves which, beside phytochemicals responsible for the burning (acetylcholine, histamine, 5-hydroxytryptamine, leukotrienes and formic acid), also contain phenolic acids, flavonoids, fatty acids, terpenes and protein. Among the phenolic acids, chlorogenic, caffeoylmalic, caffeic, gallic and quinic are contained in nettle. *U. dioica* is a rich source of other bioactive compounds: kaempferol, isorhamnetin, quercetin and its derivative, as well as patuletin and its glycosidic derivatives [97]. Essential oil of *U. dioica* contains more than 40 compounds, of which 70% are carvacrol, carvone, naphthalene, (E)-anethole, hexahydrofarnesyl acetone, (E)-β-ionone and phytol [98].

Health benefits Common nettle has been known and used as a medicinal plant since ages. The plant is considered more as a weed than an herb, but at the same time, it is also characterized by a number of pro-health properties, for which it is appreciated. Regular consumption of teas, juices and extracts of *U. dioica* shows immunostimulatory, anti-inflammatory, anticarcinogenic and antioxidant activities. Extracts obtained from different parts of the plant are used in many parts of the world. Leaves are recommended as a nutritional tonic, in the treatment of rheumatic conditions, lower urinary tract infections and for the treatment of allergies. What is more, they are used as expectorants, purgatives, diuretics and haemostatics and for the treatment of eczema, haemorrhoids, bronchitis and cancer. Bioactive compounds contained in nettle extracts may enhance selective gastric functions and protect the gastric mucosa from chemical-induced damage. Roots of the nettle, in the form of extracts, are used to reduce complaints associated with prostate hyperplasia [97].

Antimicrobial activity Phytochemicals contained in the nettle show a broad spectrum of antibacterial activity. Most commonly consumed in the form of teas and infusions, aqueous extract shows antibacterial activity against: *Micrococcus luteus, Proteus mirabilis, Citrobacter koseri, S. aureus, S. pyogenes, S. epidermidis, Streptococcus pneumoniae, Enterobacter aerogenes, E. coli*, as well as antifungal activity against *C. albicans* [99–102]. *U. dioica* extracts also show bactericidal properties against *Acinetobacter calcoaceticus, Bacillus cereus, Bacillus spizizenii, Vibrio* *parahaemolyticus* and *Klebsiella pneumonia*. What is more, the activity of nettle extract can be compared with antibiotics: miconazole, amoxicillin and ofloxacin [101]. Furthermore, the extract of *U. dioica* also shows inhibitory activities on the *Asaia* spp.—novel beverage spoilage bacteria inhabiting fruit-flavoured mineral water and isotonic drinks. Therefore, nettle is considered as an unconventional additive to this products as a food preservative [103].

4.6. Elder flowers (*S. nigra*)

Description Elderberry (*S. nigra*), also called black or common, is a deciduous shrub reaching up to 6 m high, belonging to the *Adoxaceae* family. The plant is native to sunlight-exposed areas of Asia, Africa, North America and Europe. Every summer, its flowering occurs in the form of white hermaphrodite flowers in large corymbs. The plant is highly valued mainly for the fruit, which will be described in Section 5.4 of this chapter. However, elderflower extracts are used in the beverage industry and as food flavouring as well as in alternative medicine [104].

Bioactive compounds Flowers of elderberry are a rich source of phenolic compounds, containing 10 times more flavonols than fruit. In addition to flavonols (kaempferol-3-glucoside, kaempferol-3-rutinoside, quercetin-3-glucoside, quercetin-3-rhamnoside and flavones such as apigenin), they contain derivatives of caffeic and p-coumaric acids, rutin, lupeol, β-sitosterol, tannic acid and choline [104].

Health benefits The extract of elderberry flowers has been used in traditional medicine for treatment of influenza A and B, colds, as well as an agent against the H1N1 virus. It has been documented that elderberry flower extract can be used as an agent preventing the viral adhesion of host cell receptors. What is more, the positive impact of elderberry flowers was observed in the studies on diabetes, vascular system and obesity. It has been documented that elderberry flower extract decreased fat accumulation, and hence body weight, improving the body mass index. Elderberry may have a role in the prevention and treatment of diabetes—elderberry extract may be responsible for the increase of glucose uptake and increase in the insulin production [105–107].

Antimicrobial activity Despite the fact that in folk medicine elderberry flowers are the raw material for many kinds of ailments, their antimicrobial properties are barely examined. In the studies of Mohammadsadeghi, et al. and Hearst, et al., elderberry extracts exhibit strong antibacterial activity against both gram-negative and gram-positive bacteria such us *P. aeruginosa*, *E. coli*, *Salmonella* spp., *S. aureus* and *B. cereus*. It was also demonstrated that *S. nigra* inhibits the growth of the yeast *C. albicans* [108, 109]. Our study on the antibacterial activity of the elderberry flower ethanolic extract against *A. lannensis* and *A. bogorensis* showed that the tested extract had the strongest activity against these strains. Moreover, the extract exhibited strong antiadhesive properties against all of the tested strains of *Asaia* spp. It was speculated that this broad spectrum of antibacterial activities may be the result of high content of flavonols [103].

4.7. Cinnamon bark (*C. zeylanicum*)

Description Cinnamon (*C. zeylanicum*) belonging to the family *Lauraceae* is a tropical evergreen tree originating from areas of tropical climate in India, Sri Lanka and Burma. Due to that, it is cultivated in Asia, Africa, South and Central America. Tree reaching 10 m in height mostly does not exceed 3 m. The spice is the bark of the tree, which is collected 2–3 times a year during the wet season. The bark is cut into pieces of 3 m in length and approximately 2.5 cm in diameter. Then, the bark is subjected to a short fermentation and next the removal of external and internal phloem. After that, the obtained bark is dried in the sun. During this process, it gains the characteristic yellow-brown colour. Cinnamon is widely used in ethnomedicine and as a flavouring for foods all around the world [110].

Bioactive compounds It is believed that cinnamon bark oil contains more than 70 phytochemicals comprising of: monoterpenes, oxygenated monoterpenes, sesquiterpenes, phenylpropanoids and benzenoids. The main components of the cinnamon bark extract are (E)-cinnamaldehyde, followed by (E)-cinnamyl alcohol, terpinen-4-ol, eugenol, linalool, (E)-cinnamyl acetate, o-pinene, limonene, 1,8-cineole, coumarin and β-caryophyllene [111]. On the other hand, *C. zeylanicum* bark water and ethanolic extracts can be sources of trimeric, and higher oligomeric proanthocyanidins, protocatechuic acid, as well as caffeic, chlorogenic and cinnamic acids. Additionally, it is a source of cinnamtannin B1, urolignoside, rutin, quercetin-3-rhamnopyranoside, kaempferol, procyanidin B1, apigenin and cinnamaldehyde. The last of these compounds is a highly electronegative phytochemical with many biological activities [103, 112, 113].

Health benefits The phytochemical constituents of *C. zeylanicum* may help with many different health problems. In folk medicine, it is used as a therapeutic agent against influenza, urinary tract inflammation, and as an antimicrobial agent. Aqueous extracts from the bark of *C. zeylanicum* can be responsible for the loss of weight, reducing blood glucose levels and LDL as well as increasing HDL cholesterol. It is known that cinnamon shows anti-inflammatory and antigastric activities, e.g. inhibiting gastric haemorrhagic lesions. In addition, the extract may show hepatoprotective effects. The extract also shows beneficial effects against neuropathy and nephropathy. Moreover, this extract also shows beneficial effects in Alzheimer's disease, inhibiting tau proteins aggregation and filament formation. What is more, it is believed that cinnamaldehyde extract can influence collagen biosynthesis regulating mRNA and type I collagen protein expression and thus may be useful in antiaging treatment [114].

Antimicrobial activity Both extracts and essential oils obtained from cinnamon are characterized by strong activity against broad spectrum of food poisoning microorganism, food spoilage microorganisms and human pathogens. Antibacterial activities of *C. zeylanicum* EOs have been documented against: *Acinetobacter* spp., *Clostridium perfringens, E. coli, K. pneumonia, P. aeruginosa, Salmonella typhi, B. subtilis, S. aureus, Streptococcus faecalis, S. pyogenes, Streptococcus agalactiae, S. pneumonia, Yersinia enterocolitica, Helicobacter pylori, Mycobacterium tuberculosis* and *Haemophilus influenza.* On the other hand, cinnamon extract shows inhibitory effect on *B. cereus, Bacillus coaguiaris, B. subtilis, P. aeruginosa, L. monocytogenes, Acinetobacter baumannii, E. cloacae, S. aureus, E. coli,* and food spoilage *A. lannensis* and *A. bogorensis*. What is more, essential oil of cinnamon shows antifungal activity against *Trichophyton* species (*Trichophyton mentagrophytes, T. tonsurans, T. rubrum*), *Microsporum* species (*M. canis, Microsporum gypseum, Microsporum audouinii*), *Candida* species (*C. albicans, Candida glabrata, C. parapsilosis, C. tropicalis*) as well as *Aspergillus* species (*A. fumigates, A. flavus, A. niger, A. terreus, A. ochraceus, A. nididans*) [103, 114–120].

4.8. Cloves (*S. aromaticum*)

Description Other aromatic spices popular all over the world are cloves (*S. aromaticum*), dried flower buds of clove tree belonging to the *Myrtaceae* family. The name of the spice comes from '*clavus*' which means nail. The clove tree is an evergreen tropical plant, native to Indonesia, but nowadays, the largest producer of cloves is Tanzania, then, Madagascar, Sri Lanka, Kenya, and the Seychelles. The dark colour of cloves is the result of its drying and fumigation, while the characteristic aroma is given by high concentration of eugenol. Due to the presence of essential oils in cloves, they have been used in India and China for over 2000 years and its oldest medicinal use was in China in around 240 BC [121].

Bioactive compounds *S. aromaticum* is characterized as one of the major plant sources of phenolic compounds, including flavonoids (kaempferol, quercetin and its derivatives), hydroxybenzoic acids and hydroxycinnamic acids (caffeic, gallic, ferulic, ellagic and salicylic acids and their derivatives such as hydrolyzable tannins). Essential oils in the clove flower buds are present in the concentration of 18%. Eugenol is the main bioactive compound of clove EOs and can be found in the concentration of 15 g per 100 g of the cloves. Simultaneously, the chemical compounds constitute 89% of the *S. aromaticum* EOs and are followed by eugenol acetate, β-*caryophyllene*, α-humulene, β-pinene, limonene, farnesol, benzaldehyde and 2-heptanone [122].

Health benefits Clove has been traditionally characterized as a medicinal plant with a wide range of pharmacological effects for centuries. Due to the high content of eugenol, oils and extracts are used as an antiviral agent, in the treatment of hiccups, and as antibacterial and antifungal agents. Since the thirteenth century, clove has been used as an analgesic agent. It was also reported that it may be used as an anaesthetic agent, acaricide and anticonvulsant. It is considered that the clove bioactive compounds possess great potential for pharmaceutical, cosmetic, food and agricultural applications [122]. It has been described that clove tea promotes the flow of saliva and gastric juices and can be used for stomach pain and gasses as well as for nausea and vomiting. Externally, EOs bring relief in muscle cramps, nerve conditions, chronic rheumatism, lumbago and toothache [123].

Antimicrobial activity Due to the characteristic aroma and antimicrobial activities, *S. aromaticum* can be used as a food flavouring and preservative. It has been documented that water extract shows bactericidal effect against food-borne pathogens such as *E. coli, S. aureus, B. cereus, L. monocytogenes, Listeria innocua and Salmonella enterica,* while eugenol from the cloves can inhibit the growth of *H. pylori* [124–126]. Clove oil is also described as an antifungal agent against: *Mucor* spp., *M. gypseum, M. canis, Fusarium monoliforme, T. rubrum, Aspergillus* strains, *Fusarium oxysporum* as well as against dermatophytes: *Trichophyton strains* (*T. mentagrophytes, T. rubrum*) [127, 128]. Therefore, clove extracts and essential oils show a great potential as food additives. They are a natural, effective antimicrobial agent and at the same time, give the food, a characteristic aroma which is preferred by the consumer.

4.9. Licorice (*G. glabra*)

Description Licorice is the root of *G. glabra* (*Leguminosae* family)—herbaceous perennial plant, growing up to 1 m high. The herb is generally native to southern Europe and parts of Asia, but nowadays, it can be found in India, Iran, Italy, Afghanistan, China, Pakistan, Iraq, Turkey as well as in England. The herb has been known in Chinese medicine and is believed to help to harmonize. Due to its medicinal properties and sweet taste, licorice has been used as a sweetening and flavouring agent in beverages, candies, tobacco and folk medicine [129].

Chemical composition *Glycyrrhiza* species has been documented as a source of 400 phytochemicals, from which triterpenoid saponins and flavonoids (licochalcone B, licochalcone A, echinatin, glycycoumarin and glyurallin B) are reported to be the main chemical compounds [129, 130]. The ethanolic extract of licorice root may contain genistein, glabrol, licochalcone C, as well as licorice glycoside A, licorice saponin A3 and glycyrrhizin [103] (**Figure 6**).

Health benefits Licorice has been reported as a source of biologically active phytochemicals. It is a well-known medicament in Chinese traditional medicine and is gaining popularity in other regions of the world. Licorice has been found to exhibit beneficial properties for the human organism. *Gracilimus radix* has been reported to possess antioxidative, antitumour (especially licochalcone A), antiviral, anti-inflammatory and immunity-stimulating properties. Glabridin, licoricidin and licorisoflavan A are mainly responsible for the anti-inflammatory and antioxidative activities. Moreover, glycyrrhizin from licorice has been shown to be responsible for the hepatic protective and antiulcer effects [131, 132].

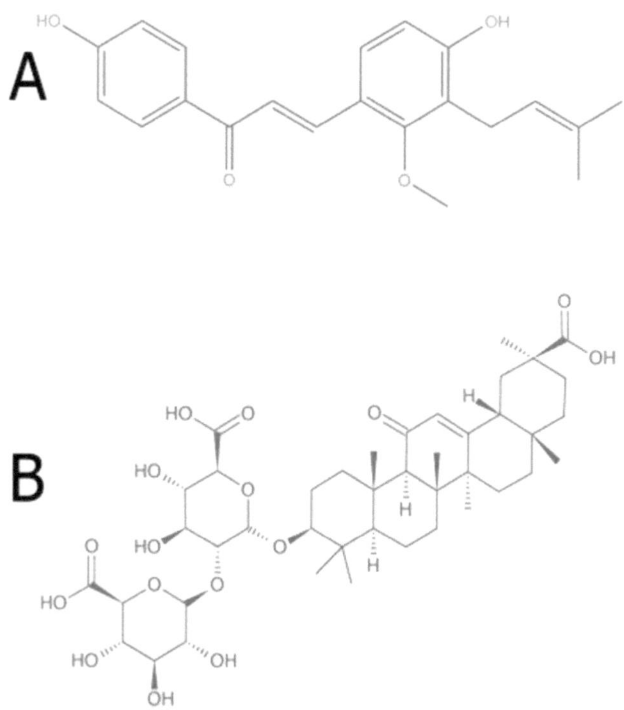

Figure 6. Chemical structure of licochalcone C (A) and glycyrrhizin (B).

Antimicrobial activity It has been noted that the extract of aerial parts of *G. glabra* exhibited antibacterial activity against *S. aureus, E. coli, B. subtilis, E. aerogenes* and *K. pneumoniae*. What is more, glabridin, glabrene and licochalcone A have been noted as antimicrobial compounds against *H. pylori*, while glycyrrhizol A showed antibacterial activity against *Streptococcus mutans* as well as *Mycobacteria* (*M. tuberculosis, Mycobacteria bovis*) species and *Legionella* (*Legionella pneumophila, Legionella bozemanii, Legionella dumoffii, Legionella longbeachae* and *Legionella wadsworthii*) species [133, 134]. What is more, probably due to the content of saponins, the extract from *G. glabra* showed antiadhesive activity against *A. lannensis* and *A. bogorensis* to food-packaging materials [103].

5. Fruit juices

In addition to essential nutrients, food also provides other health-promoting, bioactive compounds. It has been documented that a diet, and its certain components, plays a crucial role in the prevention of diseases and the treatment of others. A group of food products that is considered to be strongly associated and responsible for the reduction of the risk of cancer, diabetes, cardiovascular disease, ageing and many other are fruit. They are the major source of vitamins, micronutrients, macronutrients and secondary metabolites [135]. Secondary metabolites are a particularly important group demonstrating a broad spectrum of biological activities. Phenolic phytochemicals in plants are responsible for the protection against biological stresses in response to pathogenic attack and environmental conditions such as prolonged UV exposure [136]. There are numerous types of phytochemicals occurring in a relatively small amount in fruit. However, a group of secondary metabolites which have the health-promoting agents and are widespread in fruit are polyphenols, the most common of which are anthocyanins—responsible for red to blue and purple-black colours in fruit. Other health-benefiting compounds contained in fruit are tocopherols, glucosinolates, organosulphur compounds, sterols, stilbenes and tannins [137].

Particularly important, from the point of view of health-promoting properties and the possibility of use as additives (preservatives, colourings), are berries. They belong to the widespread family of fruits occurring in Europe, USA, Canada as well as countries of South America (Brazil, Colombia, Argentina, Paraguay and Uruguay). Main genera of berry fruits are *Vaccinium* spp.: *V. corymbosum* (blueberry), *V. myrtillus* (bilberry), *V. macrocarpon* (cranberry); *Fragaria* spp.: *F. virginiana* (strawberry); *Rubus*: *R. idaeus* (raspberry), *Rubus fruticosus* (blackberry), *Rubus ursinus* × *R. idaeus* (boysenberry); *Ribes* spp.: *Rubus rubrum* (red currant), *R. nigrum* (black currant); *Cornus* spp.: *C. mas* (cornelian cherry); *Aronia* spp.: *A. melanocarpa* (chokeberry), *Sambucus* spp.: *S. nigra* (elderberry). On the other hand, more tropical berries such as *E. oleracea* (açaí), *Eugenia uniflora* (pitanga), *Myrciaria cauliflora* (jabuticaba) and *Myrciaria dubia* (camu-camu) have also been known as the berries characterized by a high concentration of bioactive compounds [137, 138]. These fruits are rich in flavonoids (flavan-3-ols, flavonols, anthocyanins and procyanidins) and phenolic acids that possess antioxidant activities. Berries and their bioactive compounds generally reduce the incidence and mortality of cancer, cardiovascular diseases, and other diseases caused by oxidative stress, as well as coronary heart disease and cardiac stroke. The consumption of berries also results in amelioration of human ailments such as disorders in neuronal communication, inflammatory responses as well as improved memory in age [139]. Due to climate conditions, fresh berries can be consumed for several months. However, large portion of that is consumed in the form of juices, beverages, frozen products, wines and jams. What is more, due to the variety of bioactive compounds, which are characterized by beneficial activities on the health of consumers, antimicrobial properties as well as their characteristic colour, these fruits can certainly serve as a valuable additive to food products [140].

5.1. Cranberry (*V. macrocarpon*)

Description Cranberry or American cranberry (*V. macrocarpon* or *Oxycoccus macrocarpus*) belonging to the family *Ericaceae* is an evergreen dwarf shrubs native to North America and cultivated mainly throughout the northern United States and Canada. In Europe, cranberry may refer to *Vaccinium oxycoccos*, which is cultivated in central and north Europe. The fruits and leaves of the European cranberry are smaller and are refreshing, sharp and acidic in flavour, while American cranberry is slightly apple-like. The name 'cranberry' derives from 'craneberry' named by early European settlers in America, who compared small pink or red blossoms to head and bill of a 'crane'. The fruit are mainly consumed fresh, as concentrates, which have various value-added applications and juices [136, 141].

Bioactive compounds Cranberries have been recognized as a source of cyanidin-3-glucoside, cyanidin-3-galactoside, cyanidin-3-arabinoside, peonidin-3-glucoside, peonidin-3-galactoside, peonidin-3-arabinoside, delphinidin-3-glucoside, petunidin-3-glucoside and malvidin-3-glucoside, with a dominant concentration of peonidin-3-glucoside and cyanidin-3-glucoside. Among the phenolic acids, cranberry juice contains ellagic acid, ferulic acid, gallic acids, chlorogenic acids and neochlorogenic acids. Sour taste of fruit is caused by a high content of organic acids such as citric, malic and quinic. Cranberry also contains terpenes such as ursolic acid derivatives: cis-3-O-p-hydroxycinnamoyl ursolic acid and trans-3-O-p-hydroxycinnamoyl ursolic acid, as well as iridoid (monotropein) and coumaroyl iridoid glycosides. An analysis of the fractionation of cranberry juice guided by a bacterial antiadhesive assay revealed the presence of two new coumaroyl iridoid glycosides. What is more, these fruits are one of the sources of type A proanthocyanidin which is considered to be a bacterial antiadhesive agent [140, 141].

Health benefits Cranberries and cranberry products show various health-promoting properties. These berries are considered as one of the most popular and most effective natural treatments for urinary tract infections caused by uropathogenic strains of *E. coli*. It has also been described that phenolic acids and flavonoids from *V. macrocarpon* reduce oxidation of LDL, and thus the atherosclerotic process and cardiovascular disease. What is more, bioactive compounds from the juices of cranberry are able to modulate the induction of ODC (ornithine decarboxylase) and quinone reductase, which are responsible for tumour cell proliferation. Antitumor activity can also result from the inhibition of cancer cell proliferation. It is know that cranberry juice shows such activity against breast, prostate, lung, and leukaemia cells [136]. What is more, phenolic compounds from *V. macrocarpon* exhibit antiviral (against influenza A virus and type-1 herpes simplex virus), antimutagenic, antiangiogenic, anti-inflammatory and antioxidant activities [142].

Antimicrobial activity In addition to the antiadhesion properties of cranberry juice against uropathogenic strains of *E. coli,* the compounds contained in these fruits inhibit the attachment of *H. pylori* to human erythrocytes and human gastric mucous. Twenty-five percent of cranberry juice inhibited adsorption of oral pathogens *Streptococcus sobrinus* cells to saliva-coated hydroxyapatite beads in 10 seconds of contact time [143]. The proanthocyanidins from cranberry also showed inhibitory activity against gram-positive *S. epidermidis, S. aureus* and *Staphylococcus saprophyticus* [144]. It was documented that cranberry concentrate showed noticeable antimicrobial effect against *E. coli* O157:H7, *L. monocytogenes* and *S. typhimurium* [145]. Additionally, 10% cranberry juice decreased the adhesion of acetic acid bacteria *A. bogorensis* to food-packaging material [141]. Cranberry can also be an interesting candidate for natural preservation against fungal growth. Ermis et al. noted that a concentrate of cranberry juice can inhibit the growth of *Penicillium* spp., *Absidia glauca, Penicillium brevicompactum, Saccharomyces cerevisiae* and *Zygosaccharomyces bailii* [146]. It is believed that the juice can increase the microbiological safety of beverages and at the same time constitute as an alternative to the chemicals applied as food colourings.

5.2. Bilberry (*V. myrtillus*)

Description Bilberry (*V. myrtillus,* family *Ericaceae*) are Eurasian shrubs bearing small, nearly black berries. Bilberries are characteristic for North, East and Central Europe as well as Russia where they occur in forests. They are found in acidic, nutrient-poor soils, despite the fact that commercial cultivation is hard and berries are collected from natural environment. The name comes from Danish word 'bølle'. Despite the fact that the name is also used for a blueberry (*V. corymbosum*), there are a few differences between these two. Bilberries produce single or paired berries on the bush, while blueberries are gathered in clusters. Due to the significant amount of anthocyanins, *V. myrtillus* pulp is dark purple, even black, while the pulp of *V. corymbosum* is light green. What is more, shrubs of blueberry have more evergreen leaves [147].

Bioactive compounds The fruits of bilberries are one of the richest natural sources of anthocyanins, but they also contain flavanols, flavonols, phenolic acids and stilbenes. Among the anthocyanins commonly isolated from this material are: glucoside, galactosides and arabinosides of: delphinidin cyanidin petunidin peonidin and malvidin. Additionally, flavanol monomers (catechin, epicatechin), flavonols (quercetin, myricetin) as well as phenolic acids (chlorogenic, caffeic, ferulic, p-coumaric, ellagic, gallic acids) are detected. Bilberry triterpenoids consist of α- and β-amyrin, taraxasterol α-amyrenone and β-amyrenone, campesterol, citrostadienol, stigmasterol, sitostanol, cycloartenol and friedelin [148]. What is more, the presence of A type procyanidin trimer in the bilberry juice was also detected. The chemical composition of the bilberry fruits can be strongly dependent on the plant genotypes and environmental conditions [20, 147, 148].

Health benefits Due to the high content of anthocyanins, bilberries are considered to be one of the richest sources of antioxidants. *V. myrtillus* has been used in folk medicine since the

Middle Ages [148]. Fruits, juices and concentrates were all used as an antidiabetic, astringent and antiseptic agents as well as in a treatment for diarrhoea. Extracts of bilberry are documented as a source of bioactive compounds reducing illnesses such Parkinsonism, cancer and lungs diseases as well as Alzheimer's dementia [20, 149, 150]. In view of the improvement of elasticity and permeability of the capillary vessels of the eyeball, supplements with bilberry are used for the treatment of blood vessels disorders, thus improving microcirculation of the blood and vision. What is more, regular consumption may delay aging. It is also recognized that bilberries have a potential in the preventive management of type 2 diabetes, cardiovascular diseases, inflammation and hypertension [151].

Antimicrobial activity Bilberry phenolics have been reported to show antimicrobial effects against human pathogens, including *Salmonella* spp., *S. aureus*, *B. cereus* and *S. epidermidis*. Bioactive compounds from berries are generally able to inhibit *H. pylori*, *E. coli*, *Citrobacter freundii* and *Enterococcus faecalis*. Moreover, the phenolic compounds such as anthocyanins and flavonols can inhibit the growth of *Salmonella* spp. and *E.coli*, while tannins exhibit strong antimicrobial effect against *C. perfringens*, *Klebsiella* spp., and *Proteus* spp. [152–155]. Additionally, it has been shown that the juice of bilberry has a strong antiadhesive effect against gram-negative, beverage spoilage bacteria belonging to the genus *Asaia* [20].

5.3. Black currant (*R. nigrum*)

Description Black currant (*R. nigrum*) is a shrub in the family *Grossulariaceae* growing up to 2 m high in various parts of the world with temperate climate. *R. nigrum* is native to central and northern Europe and northern Asia, cultivated commercially and domestically. The plant can grow well on sandy and forest soils, as it does not tolerate both waterlogged grounds and droughts. Fruits, flowers, leaves bark and roots are strongly aromatic and have been used in traditional medicine. *R. nigrum* has been described as a garden plant in Russia in the eleventh century. It has been cultivated in Europe since the seventeenth century. It is worth mentioning that black currant shrubs were widely cultivated in the United Kingdom during World War II. Syrups obtained from the fruit were a source of vitamin C and were distributed free of charge to children under the age of two [156].

Bioactive compounds Berries of black currant contain many bioactive compounds showing benefits to the human health. Similar to the bilberry, *R. nigrum* are characterized by a high content of anthocyanins, which give them a strong purple colour. The source of the anthocyanins is the skin, containing: cyanidin-3-glucoside, cyanidin-3-rutinoside, delphinidin-3-glucoside and delphinidin-3-rutinosid, of which delphinidins are the main compounds. The main representative of the phenolic acids group are chlorogenic and neochlorogenic acids. Black currant fruits also contain: flavonols (myricetin and quercetin glycosides) as well as catechins (epigallocatechin, gallocatechin, epicatechin and epigallocatechin gallate) [156, 157].

Health benefits Black currant has recently been labelled as the so-called *super fruit* and is believed to possess several health benefits. Traditionally, juices and extracts of black currant fruits have been used as a protection against viral and bacterial infections. It was documented that it inhibited the influenza virus (IFV-A and IFV-B), the herpes simplex virus (HSV-1 and

HSV-2), inhibiting the protein synthesis of infected cells, as well as viruses associated with upper respiratory tract such as respiratory syncytial virus (RSV) and adenovirus (AdV). Bioactive compounds of the *R. nigrum* fruits show protective activities against neuronal damages. Anthocyanins decrease blood pressure, reduce muscle fatigue as well as enhance peripheral circulation. What is more, the juice shows strong activities against cancer cells proliferation [158]. It was described that consumption of *R. nigrum* can inhibit the growth of different cancer cells: Caco-2, human breast cancer cell lines (MDA-MB-231, MCF-7), human gastric carcinoma (AGS), human prostate cancer (PC-3) and human colon adenocarcinoma (HT-29). Moreover, *R. nigrum* extracts are rich in anthocyanin and exhibit antioxidant, anti-inflammatory and immunostimulatory properties [159, 160].

Antimicrobial activity Strong inhibition was observed in relation to *Serratia marcescens*, *B. subtilis*, *E. coli*, as well as *S. typhimurium*, *Campylobacter jejuni* and *Streptococcus pneumonie*—bacterium responsible for severe meningitis and pneumonia in infants [153, 161–163]. It is considered that blackcurrant juices are generally more efficient against gram-positive than against gram-negative bacteria. Anticandidal activity against *C. albicans*, *Caesalpinia pulcherrima*, *C. krusei* and *C. lusitaniae* was also observed [164]. In addition to growth inhibition, the juice of black currant possesses anti-adhesive activity against *Asaia* species [20]. Black currant extracts show inhibition of microbial growth and adsorption and can be easily used as a natural preservative in food products.

5.4. Elderberry (*S. nigra*)

Description Elderberry (*S. nigra*) is a shrub reaching up to 6 m high belonging to the *Adoxaceae* family. The plant is native to sunlight-exposed areas of Asia, Africa, North America and Europe [104]. It is believed that early settlers brought elderberry seeds from Europe (*S. nigra*) and the plant became naturalized in North America (*Sambucus canadensis*). The name *Sambucus* probably derives from the Greek '*sambuke*' or the Latin '*sambuca*', referring to a kind of flute which has been made out of twigs. In the beginning, elder berries are oblong, compact and green, and after 6–8 weeks, they gradually enlarge and change colour from red to purple and black. Ripe fruits may range from 5 to 7 mm in diameter, weighing approximately 50–130 mg [165].

Bioactive compounds Similar to other dark berries, anthocyanins are mainly responsible for the colour of the elderberry. In addition, they contain flavonols, phenolic acids and proanthocyanidins. *S. nigrum* anthocyanins are cyanidin derivatives (e.g. cyanidin-3-sambubioside-5-glucoside), pelargonidin derivatives (pelargonidin-3-sambubioside) and delphinidin derivatives (delfinidine-3-rutinoside). Due to the presence of acylated form of anthocyanins, American elderberries show greater diversity in the composition. Elderberry fruits are also one of the richest sources of phenolic acids: chlorogenic, crypto-chlorogenic, neochlorogenic acids and ellagic acid, but their content is less diverse than in those of the flowers of elderberry. Among flavonols, the fruits contain quercetin, kaempferol and isorhamnetin as well as their glycosylated forms (rutin and glucose). Proanthocyanidins occur in elderberries in a relatively small concentration, few times lower than in chokeberry and black currant [107].

Health benefits The use of elderberry fruits as well as other parts of the plant: flowers, leaves, roots and bark has been known since ages. The first report mentioning the use of the elderberry, *De Materia Medica* comes from Ancient Rome. In American and European cultures,

Sambucus spp. has been used in folk medicine in treating respiratory diseases (influenza, colds and catarrh), and as a diaphoretic, diuretic, laxative and anti-inflammatory, natural bioactive agent. It has also been used for swelling, haemorrhoids, rheumatic symptoms, toothaches, kidney and eye problems, as well as in hepatitis and dyspepsia. What is more, the *S. nigrum* extracts can exhibit immunostimulatory activities (by the stimulation of the production of IL-1β, IL-6, IL-8 and IL-10 as well as TNF-α), is responsible for the reduction of glycaemia and may reduce blood pressure [107]. However, the fruits are most known for their antiviral activity, including against the H1N1 virus [166].

Antimicrobial activity Antimicrobial properties of extracts from different parts of elderberry (*S. nigrum*) have been documented against *B. cereus, Serratia marcescens, E. coli, S. aureus, P. aeruginosa Salmonella* spp. as well as *B. subtilis, B. megaterium*, and yeasts: *D. hansenii, Z. rouxii, Rhodotorula rubra, Candida shehatae* and *C. tropicalis* [107, 109]. Commercially standardized extracts of elderberry such as 'Rubini' showed antimicrobial activities against human pathogens: *S. pyogenes* and *Branhamella catarrhalis* [167]. Extract from elderberry showed inhibitory effect of the growth of *Mycoplasma mycoides* subsp. *capri, E. coli, B. subtilis* [168], and clinical strains of *H. pylori* [169].

5.5. Cornelian cherry (*C. mas*)

Description The cornelian cherry (*C. mas*, family *Cornaceae*) is considered to be a high deciduous shrub or a small tree growing from 5 to 8 m. The cornelian cherry is a rare plant occurring in Europe (Belgium, Germany, the Czech Republic, Slovakia, UK as well as Turkey which is the main producer of these fruits). The plant is characterized by its extraordinary tolerance to environmental conditions. There is a saying 'healthy as the Cornelian cherry'. The fruits are long, cherry-like with sour taste [170].

Bioactive compounds *C. mas* is particularly rich in ascorbic acids and anthocyanins such as delphinidin galactopyranoside, delphinidin rutinoside, delphinidin glucoside, cyanidin glucoside, cyanidin rutinoside, cyanidin galactopyranoside, pelargonidin galactopyranoside and pelargonidin-3-glucoside. What is more, phenolic acids (gallic, ellagic, chlorogenic), as well as (+)-catechin, (-)-epicatechin, procyanidin B2, luteolin-3-glucoside, kaempferol-3-glucoside occur as the bioactive composition of the *C. mas* fruits [170–172].

Health benefits Fresh fruits of Cornelian cherry from Greece are characterized by one of the strongest antioxidant activities compared to other fruit from this region. Further, *C. mas* are known as a source of phytochemicals showing antihistamine, antiallergic, antimalarial as well as antidiabetic potential. They have also been used as a source of bioactive compounds showing beneficial effect on liver and kidney. In Asian countries, cornelian cherry has been used as an antidiabetic agent. The high level of antioxidants in these fruits makes them a candidate for the prevention or treatment of neurological diseases [172–174].

Antimicrobial activity Extracts from fresh *C. mas* fruits have been shown to possess strong antibacterial activity against both gram-positive and gram-negative bacteria: *S. aureus* and *P. aeruginosa* [175]. Wide spectrum of antibacterial and anticandidal activities of both juice of *C. mas* and extract of their leaves has been documented in the article of Milenković-Andjelković et al. The author noted inhibition in growth of: *C. perfringens, B. cereus, S. aureus, L. monocytogenes,*

Sarcina lutea, M. flavus, E. coli, P. aeruginosa, Salmonella enteritidis, Shigella sonnei, K. pneumoniae, P. vulgaris as well as *C. albicans* [171]. However, further antimicrobial properties of this plant are being studied and raising great interest as an additive to food products.

5.6. Açaí (*E. oleracea*)

Description Açaí (*Euterpe oleracia)* is a species of palm tree belonging to the family *Arecaceae*, cultivated in South America, the Amazonian flood lands, Brazil and Columbia. In recent years, an increasing demand for both the heart of the palm trees, as well as its fruits has been observed. The trees are tall, reaching over 25 m. The fruits are purple black, about 10 mm in diameter, with flavour similar to raspberries. Visually, they are similar to grapes, but they are much smaller and are produced in branched panicles of 900 fruits. Due to the low sugar value, the fruits are not sweet but contain a high amount of dietary fibres [137].

Bioactive compounds It is recognized that these fruits are one of the richest and most diverse sources of bioactive compounds. Among the anthocyanins, açaí berries contain cyanidin-3-glucoside, cyanidin-3-rutinoside, cyanidin-3-acetylhexose, cyanidin-3-arabinoside, cyanidin-3-sambubioside, peonidin-3-rutinoside, peonidin-3-glucoside and pelargonidin-3-glucoside. In addition, they also contain flavonoids: apigenin diglucoside, homoorientin, orientin, taxifolin deoxyhexose, taxifolin-3-rhamnoside, isovitexin, velutin, scoparin, as well as catechin and epicatechin. Additionally, the presence of procyanidin dimers and trimers was also noted. An important part of the phenolic components in the fruits is phenolic acids: protocatechuic, p-hydroxybenzoic, vanillic, syringic, ferulic, gallic, benzoic, p-coumaric and ellagic. It is hardly surprising that *E. oleracea* fruits are considered to be one of the most important sources of health benefiting phytochemicals and recognized as '*super food*' [137].

Health benefits Due to the richness in bioactive compounds, they are a promising health-benefiting food product or food additive. It is believed that açaí could play an important role in the prevention of cancer, cardiovascular diseases show antioxidant action in relation to human endothelial cells and show effects on epigenetics modulators, such as microRNAs. The juice when drunk regularly can have positive effects on blood lipid levels and can protect the heart from coronary heart disease, as well as reduce probability of type 2 diabetes. Further, it was noted that the extract of *E. oleracea* is responsible for the reduction of total cholesterol levels and can be helpful in weight reduction and maintaining healthy weight. Due to the high phenolic content and dietary fibres, these fruits can be used in the treatment of digestive problems. Additionally, regular consumption of *E. oleracea* can result in increased focus and memory improvement, as well as protect against the damaging effects of stress, and thus, it is considered to be an '*adaptogen*' [131]. Therefore, it is considered to commercialize the juice from these berries as an additive to different food products, for example beverages [176].

Antimicrobial activity Despite the fact that the berries have long been used in folk medicine as therapeutic agents that contain one of the highest concentrations of bioactive compounds, the data on its antimicrobial activities are very limited. However, the similarity in the composition of phytochemicals to other berry fruits allows us to assume that açaí may show such activity [137, 176].

6. Conclusion

Microbial food safety is a constant, global problem affecting the health of consumers. Due to the increasing resistance of microorganisms to chemicals used for the technological lines disinfection, as well as lower sensitivity to the synthetic preservatives, alternative sources of natural, bioactive, and antimicrobial compounds are needed. The information presented in our chapter shows that certain fruit, herbs and spices as well as phytochemicals, and their mixtures are characterized by strong antimicrobial and antiadhesive activities against food-borne pathogens, food spoilage bacteria, yeasts and moulds, as well as against human pathogens. Simultaneously, they have health-promoting properties and show therapeutic and preventative effects. Furthermore, from the technological point of view, essential oils, plant extracts and fruit juices may be used in food products as other additives than preservatives. Due to the essential oil content, they may be used as flavourings, while a high level of anthocyanins in berries makes them an ideal candidates to be colouring agents. There is no doubt that these natural products can be used in food production, helping to maintain stability of the products and meeting the demands of consumers in relation to natural food additives.

Author details

Hubert Antolak* and Dorota Kregiel

*Address all correspondence to: hubert.antolak@gmail.com

Lodz University of Technology, Łódz, Poland

References

[1] Fuleky G. Cultivated Plants, Primarily as Food Sources. Vol. 1. EOLSS Publishers Co Ltd. Paris, France; 2014

[2] Botanic Gardens Conservation International. Available from: https://www.bgci.org/policy/1521/ [Accessed: 15 February 2017]

[3] World Health Organization Report. Available from: http://www.who.int/mediacentre/factsheets/2003/fs134/en/ [Accessed: 15 February 2017]

[4] Carpenter S, Rigaud M, Barile M, Priest TJ, Perez L, Ferguson JB. An Interlinear Transliteration and English Translation of Portions of the Ebers Papyrus Possibly Having to Do With Diabetes Mellitus. Bard College, Annandale-on-Hudson New York, USA; 2006. p. 10

[5] Borlinghaus J, Albrecht F, Gruhlke MCH, Nwachukwu ID, Slusarenko AJ. Allicin: Chemistry and biological properties. Molecules. 2014;**9**(8):12591-12618. DOI: 10.3390/molecules190812591

[6] Hou JP. The development of Chinese herbal medicine and the Pen-ts'ao. Comparative Medicine East and West. 1977;**5**(2)**:**117-122

[7] Pandey MM, Rastogi S, Rawat AKS. Indian traditional Ayurvedic system of medicine and nutritional supplementation. Evidence-Based Complementary and Alternative Medicine. 2013:376327. DOI: 10.1155/2013/376327

[8] Iniesta I. Hippocratic Corpus. British Medical Journal. 2011;**342**. DOI: 10.1136/bmj.d688

[9] Leonti M, Casu L, Sanna F, Bonsignore L. A comparison of medicinal plant use in Sardinia and Sicily-De Materia Medica revisited? Journal of Ethnopharmacology. 2009;**2:**255-267. DOI: 10.1016/j.jep.2008.10.027

[10] Toledo-Pereyra LH. Claudius Galenus of Pergamum: Surgeon of gladiators. Father of experimental physiology. Journal of Investigative Surgery. 2002;**6**:299-301. DOI: 10.1080/08941930290086100

[11] Siraisi NG. Avicenna in Renaissance Italy: The Canon and Medical Reaching in Italian Universities After 1500. Princeton University Press, New Jersey, USA; 1987

[12] Borzelleca JF. Paracelsus: Herald of modern toxicology. Toxicological Sciences. 2000; **53**(1):2-4. DOI: 10.1093/toxsci/53.1.2

[13] Binswanger HC, Smith KR. Paracelsus and Goethe: Founding fathers of environmental health. Bulletin of the World Health Organization. 2000;**78**(9):1162-1165. DOI: 10.1590/S0042-96862000000900013

[14] Vlot AC, Dempsey DA, Klessig DF. Salicylic acid, a multifaceted hormone to combat disease. Annual Review of Phytopathology. 2009;**47:**177-206. DOI: 10.1146/annurev.phyto.050908.135202

[15] Hefferon K. Let Thy Food Be Thy Medicine: Plants and Modern Medicine. Oxford University Press, New York, USA; 2012

[16] Nobel Prize. Nobel Prizes and Laureates. Available from: https://www.nobelprize.org/nobel_prizes/medicine/laureates/2015/press.html [Accessed: 18 February 2017]

[17] Newman DJ, Cragg GM. Natural products as sources of new drugs from 1981 to 2014. Journal of Natural Products. 2016;**79**(3):629-661. DOI: 10.1021/acs.jnatprod.5b01055

[18] Bigliardia B, Galatib F. Innovation trends in the food industry: The case of functional foods. Trends in Food Science and Technology. 2013;**31**(2):118-129. DOI: 10.1016/j.tifs.2013.03.006

[19] Wildman REC. Handbook of Nutraceuticals and Functional Foods. CRC Press, Florida, USA; 2000

[20] Antolak H, Czyzowska A, Kregiel D. Black currant (*Ribes nigrum* L.) and bilberry (*Vaccinium myrtillus* L.) fruit juices inhibit adhesion of *Asaia* spp. BioMed Research International. 2016:3671306. DOI: 10.1155/2016/3671306

[21] Aloui H, Khwaldia K. Natural antimicrobial edible coatings for microbial safety and food quality enhancement. Comprehensive Reviews in Food Science and Food Safety. 2016;**15**(6):1080-1103. DOI: 10.1111/1541-4337.12226

[22] Meltzer HM. Bioactive compounds through food, nutraceuticals or pills? In: Berbhoft A. editor. A brief review on bioactive compounds in plants. Oslo, The Norwegian Academy of Science and Letters; 2010, 205-222

[23] Azmir J, Zaidul ISM, Rahman MM, Sharif KM, Mohamed A, Sahena F, Jahurul MHA, Ghafoor K, Norulaini NAN, Omar AKM. Techniques for extraction of bioactive compounds from plant materials: A review. Journal of Food Engineering. 2013;**117**:426-436. DOI: 10.1016/j.jfoodeng.2013.01.014

[24] Croteau R, Kutchan TM, Lewis NG. Natural products (secondary metabolites). In: Buchanan B, Gruissem W, Jones R, editors. Biochemistry and Molecular Biology of Plants. American Society of Plant Physiologists, New Jersey, USA; 2000. pp. 1250-1318

[25] Tiaz L, Zeiger E. Secondary metabolites and plant defense. In: Plant Physiology. Sinauer Associates, Sunderland, Massachusetts, USA; 2006. pp. 283-308

[26] Charles DJ. Antioxidant Properties of Spices, Herbs and Other Sources. Springer, New York, USA; 2013. DOI: 10.1007/978-1-4614-4310-0

[27] Robbins RJ. Phenolic acids in foods: An overview of analytical methodology. Journal of Agricultural and Food Chemistry. 2003;**51**(10):2866-2887. DOI: 10.1021/jf026182t

[28] Jakobek L, Šeruga M, Medvidović-Kosanović M, Novak I. Antioxidant activity and polyphenols of *Aronia* in comparison to other berry species. Agriculturae Conspectus Scientificus. 2007;**72**(4):301-306

[29] Nijveldt RJ, van Nood E, van Hoorn DE, Boelens PG, van Norren K, van Leeuwen PA. Flavonoids: A review of probable mechanisms of action and potential applications. American Journal of Clinical Nutrition. 2001;**74**(4):418-425

[30] Ninomiya M, Koketsu M. Minor flavonoids (chalcones, flavanones, dihydrochalcones, and aurones). In: Ramawat KG, Mérillon J-M, editors. Natural Products. Springer, New York, USA; 2013. pp. 1867-1900. DOI: 10.1007/978-3-642-22144-6_62

[31] Castañeda-Ovando A, Pacheco-Hernández M, Páez-Hernández ME, Rodríguez JA, Galán-Vidal CA. Chemical studies of anthocyanins: A review. Food Chemistry. 2009;**113**:859-871. DOI: 10.1016/j.foodchem.2008.09.001

[32] Ullah N, Abbas A, Hider F. Comparative study of alkaloids in selected medical plants of Mansehra district. International Journal of Applied Pharmaceutical and Biological Research. 2016;**1**(3):59-65

[33] Khatoon S. A novel histological approach for identification of alkaloid bearing plants. International Journal of Botany. 2017;**13**:28-36. DOI: 10.3923/ijb.2017.28.36

[34] de Sousa Falcão H, Leite JA, Barbosa-Filho JM, de Athayde-Filho PF, de Oliveira Chaves MC, Moura MD, Ferreira AL, de Almeida ABA, Souza-Brito ARM, Diniz MFFM, Batista LM. Gastric and duodenal antiulcer activity of alkaloids: A review. Molecules. 2008;**13**(12):3198-3223. DOI: 10.3390/molecules13123198

[35] Lu X, Tang K, Li P. Plant metabolic engineering strategies for the production of pharmaceutical terpenoids. Frontiers in Plant Science. 2017;**7**:1647. DOI: 10.3389/fpls.2016.01647

[36] Moses T, Papadopoulou KK, Osbourn A. Metabolic and functional diversity of saponins, biosynthetic intermediates and semi-synthetic derivatives. Critical Reviews in Biochemistry and Molecular Biology. 2014;**49**(6):439-462. DOI: 10.3109/10409238.2014.95362

[37] Kregiel D, Berlowska J, Witonska I, Antolak H, Proestos C, Babic M, Babic L, Zhang B. Saponin-based, biological-active surfactants from plants. In: Najja R, editor. Surfactants and Detergents. InTech; 2017 [in print]

[38] Güçlü-Ustündağ O, Mazza G. Saponins: Properties, applications and processing. Critical Reviews in Food Science and Nutrition. 2007;**47**(3):231-258. DOI: 10.1080/10408390600698197

[39] Du J-R, Long F-Y, Chen C. Research progress on natural triterpenoid saponins in the chemoprevention and chemotherapy of cancer. In: Bathaie SZ, Tamanoi F, editors. Natural Products and Cancer Signaling: Isoprenoids, Polyphenols and Flavonoids. The Enzymes. Vol. 36. Elsevier Inc. Cambridge, Massachusetts, USA; 2014. pp. 95-130. DOI: 10.1016/B978-0-12-802215-3.00006-9

[40] Bissinger R, Modicano P, Alzoubi K, Honisch S, Faggio C, Abed M, Lang F. Effect of saponin on erythrocytes. International Journal of Hematology. 2014;**100**(1):51-59. DOI: 10.1007/s12185-014-1605

[41] Miyakoshi M, Tamura Y, Masuda H, Mizutani K, Tanaka O, Ikeda T, Ohtani K, Kasai R, Yamasaki K. Antiyeast steroidal saponins from *Yucca schidigera* (*Mohave yucca*), a new anti-food-deteriorating agent. Journal of Natural Products. 2000;**63**(3):332-338

[42] Sucharzewska D, Stochmal A, Oleszek W. The effect of *Yucca schidigera* extract on the physical structure and on the oxidative stability of sugar-candy foam products. Lebensmittel Wissenschaft and Technologie. 2003;**36**:347-351. DOI: 10.1016/S0023-6438(03)00016-1

[43] Berlowska J, Dudkiewicz M, Kregiel D, Czyzowska A, Witonska I. Cell lysis induced by membrane-damaging detergent saponins from *Quillaja saponaria*. Enzyme and Microbial Technology. 2015;**75-76**:44-48. DOI: 10.1016/j.enzmictec.2015.04.007

[44] Peter KV. Handbook of Herbs and Spices. CRC Press, Florida, USA; 2001

[45] Mann J, Truswell AS. Essentials of Human Nutrition. Oxford University Press, New York, USA; 2002

[46] Capecka E, Mareczek A, Leja M. Antioxidant activity of fresh and dry herbs of some *Lamiaceae* species. Food Chemistry. 2005;**93**(2):223-226

[47] Tajkarimi MM, Ibrahim SA, Cliver DO. Antimicrobial herb and spice compounds in food. Food Control. 2010;**21**:1199-1218

[48] Lv J, Huang, H, Yu L, Whent M, Niu Y, Shi H, Wang TTY, Luthria D, Charles D, Yu LL. Phenolic composition and nutraceutical properties of organic and conventional cinnamon and peppermint. Food Chemistry. 2012;**132**(3):1442-1450. DOI: 10.1016/j.foodchem.2011.11.135

[49] Riachi LG, De Mariaa CAB. Peppermint antioxidants revisited. Food Chemistry. 2015;**176**(1):72-81. DOI: 10.1016/j.foodchem.2014.12.028

[50] Giamperi L, Fraternale D, Ricci D. The in vitro action of essential oils on different organisms. Journal of Essential Oil Research. 2002;**14**(4):312-318

[51] Moghaddam M, Pourbaige M, Tabar HK, Farhadi N, Hosseini SMA. Composition and antifungal activity of peppermint (*Mentha piperita*) essential oil from Iran. Journal of Essential Oil Bearing Plants. 2013;**16**(4):506-512. DOI: 10.1080/0972060X.2013.813265

[52] McKay DL, Blumberg JB. A review of the bioactivity and potential health benefits of peppermint tea (*Mentha piperita* L.). Phytotherapy Research. 2006;**20:**619-633. DOI: 10.1002/ptr.1936

[53] Fecka I, Turek S. Determination of water soluble polyphenolic compounds in commercial herbal teas from *Lamiaceae*: Peppermint, Melissa, and sage. Journal of Agricultural and Food Chemistry. 2007;**55**:10908-10917. DOI: 10.1021/jf072284d

[54] Ohara A, Matsuhisa T. Antitumor promoting activities of edible plants against okadaic acid. Food Science and Technology Research. 2012;**8**(2):158-161. DOI: 10.3136/fstr.8.158

[55] Samarth RM, Panwar M, Kumar M, Kumar A. Protective effects of *Mentha piperita* Linn on benzo[α]pyrene-induced lung carcinogenicity and mutagenicity in Swiss albino mice. Mutagenesis. 2016;**21**:61-66. DOI: 10.1093/mutage/gei075

[56] Minami M, Kita M, Nakaya T, Yamamoto T, Kuriyama H, Imanishi J. The inhibitory effect of essential oils on herpes simplex virus type-1 replication in vitro. Microbiology and Immunology. 2003;**47**(9):681-684

[57] Singh R, Shushni MAM, Belkheir A. Antibacterial and antioxidant activities of *Mentha piperita* L. Arabian Journal of Chemistry. 2015;**8**(3):322-328. DOI: 1016/j.arabjc.2011.01.019

[58] Mahboubi M, Kazempour N. Chemical composition and antimicrobial activity of peppermint (*Mentha piperita* L.) essential oil. Songklanakarin Journal of Science and Technology. 2014;**36**(1):83-87

[59] Pramila DM, Xavier R, Marimuthu K, Kathiresan S, Khoo ML, Senthilkumar M, Sathya K, Sreeramanan S. Phytochemical analysis and antimicrobial potential of methanolic leaf extract of peppermint (*Mentha piperita*: *Lamiaceae*). Journal of Medicinal Plants Research. 2012;**6**(2):331-335. DOI: 10.5897/JMPR11.1232

[60] Burt S. Essential oils: Their antibacterial properties and potential applications in foods—a review. International Journal of Food Microbiology. 2004;**94**(3):223-253. DOI: 10.1016/j.ijfoodmicro.2004.03.022

[61] Carretto CFP, Almeida RBA, Furlan MR, Jorge AOC, Junqueira JC. Antimicrobial activity of *Mentha piperita* L. against *Candida* spp. Brazilian Dental Journal. 2010;**13**(1):4-9

[62] Antolak H, Czyzowska A, Kregiel D. Anti-adhesion activity of mint (*Mentha piperita* L.) leaves extract against beverage spoilage bacteria *Asaia* spp. Biotechnology and Food Sciences. 2016;**80**(2):119-127

[63] Sandasi M, Leonard CM, Viljoen AM. The effect of five common essential oil components on *Listeria monocytogenes* biofilms. Food Control. 2008;**19**:1070-1075. DOI: 10.1016/j.foodcont.2007.11.006

[64] Sandasi M, Leonard CM, Van Vuuren SF, Viljoen AM. Peppermint (*Mentha piperita*) inhibits microbial biofilms in vitro. South African Journal of Botany. 2011;**77**:80-85. DOI: 10.1016/j.sajb.2010.05.011

[65] Saharkhiz MJ, Motamedi M, Zomorodian K, Pakshir K, Miri R, Hemyari K. Chemical composition, antifungal and antibiofilm activities of the essential oil of *Mentha piperita* L. International Scholarly Research Notices. **2012**(2012):718645. DOI: 10.5402/2012/718645

[66] Dambolena JS, Zunino MP, López AG, Rubinstein HR, Zygadlo JA, Mwangi JW, Thoithi GN, Kibwage IO, Mwalukumbi JM, Kariuki ST. Essential oils composition of *Ocimum basilicum* L. and *Ocimum gratissimum* L. from Kenya and their inhibitory effects on growth and fumonisin production by *Fusarium verticillioide*. Innovative Food Science and Emerging Technologies. 2010;**11**:410-414. DOI: 10.1016/j.ifset.2009.08.005

[67] Kwee EM, Niemeyer ED. Variations in phenolic composition and antioxidant properties among 15 basil (*Ocimum basilicum* L.) cultivars. Food Chemistry. 2011;**128**:1044-1050. DOI: 10.1016/j.foodchem.2011.04.011

[68] Lee J, Scagel CF. Chicoric acid found in basil (*Ocimum basilicum* L.) leaves. Food Chemistry. 2009;**115**:650-656. DOI: 10.1016/j.foodchem.2008.12.075

[69] Vlase L, Benedec D, Hanganu D, Damian G, Csillag I, Sevastre B, Mot AC, Silaghi-Dumitrescu R, Tilea I. Evaluation of antioxidant and antimicrobial activities and phenolic profile for *Hyssopus officinalis*, *Ocimum basilicum* and *Teucrium chamaedrys*. Molecules. 2014;**19**(5):5490-5507. DOI: 10.3390/molecules19055490

[70] Said-Al Ahl HAH, Meawad AA, Abou-Zeid EN, Ali MS. Evaluation of volatile oil and its chemical constituents of some basil varieties in Egypt. International Journal of Plant Research. 2015;**1**(3)**:**103-106

[71] Moghaddam AMD, Shayegh J, Mikaili P, Sharaf JD. Antimicrobial activity of essential oil extract of *Ocimum basilicum* L. leaves on a variety of pathogenic bacteria. Journal of Medicinal Plants Research. 2011;**5**(15)**:**3453-3456

[72] Sienkiewicz M, Łysakowska M, Pastuszka M, Bienias W, Kowalczyk E. The potential of use basil and rosemary essential oils as effective antibacterial agents. Molecules. 2013;**18**(8):9334-9351. DOI: 10.3390/molecules18089334

[73] Zhang J-W, Li SK, Wu W-J. The main chemical composition and in vitro antifungal activity of the essential oils of *Ocimum basilicum* Linn. var. pilosum (Willd.) Benth. Molecules. 2009;**14**(1):273-278. DOI: 10.3390/molecules14010273

[74] Saggiorato AG, Gaio I, Treichel H, de Oliveira D, Cichoski AJ, Cansian RL. Antifungal activity of basil essential oil (*Ocimum basilicum* L.): Evaluation *in vitro* and on an Italian-type sausage surface. Food and Bioprocess Technology. 2012;**5**:378-384. DOI: 10.1007/s11947-009-0310-z

[75] Rattanachaikunsopon P, Phumkhachorn P. Antimicrobial activity of basil (*Ocimum basilicum*) oil against *Salmonella enteritidis* in vitro and in food. Bioscience, Biotechnology, and Biochemistry. 2010;**74**(6):1200-1204. DOI: 10.1271/bbb.90939

[76] Carovic-Stanko K, Orlic S, Politeo O, Strikic F, Kolak I, Milos M, Satovic Z. Composition and antibacterial activities of essential oils of seven *Ocimum* taxa. Food Chemistry. 2010;**119**:196-201. DOI: 10.1016/j.foodchem.2009.06.010

[77] Szumny A, Figiel A, Gutiérrez-Ortíz A, Carbonell-Barrachina AA. Composition of rosemary essential oil (*Rosmarinus officinalis*) as affected by drying method. Journal of Food Engineering. 2010;**97**:253-260. DOI: 10.1016/j.jfoodeng.2009.10.019

[78] Kontogianni VG, Tomic G, Nikolic I, Nerantzaki AA, Sayyad N, Stosic-Grujicic S, Stojanovic I, Gerothanassis IP, Tzakos AG. Phytochemical profile of *Rosmarinus officinalis* and *Salvia officinalis* extracts and correlation to their antioxidant and anti-proliferative activity. Food Chemistry. 2013;**136**(1):120-129. DOI: 10.1016/j.foodchem.2012.07.091

[79] Mena P, Cirlini M, Tassotti M, Herrlinger KA, Dall'Asta C, Del Rio D. Phytochemical profiling of flavonoids, phenolic acids, terpenoids, and volatile fraction of a rosemary (*Rosmarinus officinalis* L.) extract. Molecules. 2016;**21**(11):1576. DOI: 10.3390/molecules21111576

[80] Rašković A, Milanović I, Pavlović N, Ćebović T, Vukmirović S, Mikov M. Antioxidant activity of rosemary (*Rosmarinus officinalis* L.) essential oil and its hepatoprotective potential. BMC Complementary and Alternative Medicine. 2014;**14**:225. DOI: 10.1186/1472-6882-14-225

[81] Vasile C, Sivertsvik M, Mitelut AC, Brebu MA, Stoleru E, Rosnes JT, Tănase EE, Khan W, Pamfil D, Cornea CP, Irimia A, Popa EM. Comparative analysis of the composition and active property evaluation of certain essential oils to assess their potential applications in active food packaging. Materials. 2017;**10**(1):45. DOI: 10.3390/ma10010045

[82] Moore J, Yousef M, Tsiani E. Anticancer effects of rosemary (*Rosmarinus officinalis* L.) extract and rosemary extract polyphenols. Nutrients. 2016;**8**(11):731. DOI: 10.3390/nu8110731

[83] Wang W, Li N, Luo M, Zu Y, Efferth T. Antibacterial activity and anticancer activity of *Rosmarinus officinalis* L. essential oil compared to that of its main components. Molecules. 2012;**17**:2704-2713. DOI: 10.3390/molecules17032704

[84] Bozin B, Mimica-Dukic N, Samojlik I, Jovin E. Antimicrobial and antioxidant properties of rosemary and sage (*Rosmarinus officinalis* L. and *Salvia officinalis* L., *Lamiaceae*) essential oils. Journal of Agricultural and Food Chemistry. 2007;**55**(19):7879-7885. DOI: 10.1021/jf0715323

[85] Teodoro RAR, de Barros Fernandes RV, Botrel DA, Borges SV, de Souza AU. Characterization of microencapsulated rosemary essential oil and its antimicrobial effect on fresh dough. Food and Bioprocess Technology. 2014;**7**:2560. DOI: 10.1007/s11947-014-1302-1

[86] Fachini-Queiroz FC, Kummer R, Estevão-Silva CF, de Barros Carvalho MD, Cunha JM, Grespan R, Bersani-Amado CA, Cuman RKN. Effects of thymol and carvacrol, constituents of *Thymus vulgaris* L. essential oil, on the inflammatory response. Evidence-Based Complementary and Alternative Medicine. 2012:657026. DOI: 10.1155/2012/657026

[87] Roby MHH, Sarhan MA, Selim KA-H, Khalel KI. Evaluation of antioxidant activity, total phenols and phenolic compounds in thyme (*Thymus vulgaris* L.), sage (*Salvia officinalis* L.), and marjoram (*Origanum majorana* L.) extracts. Industrial Crops and Products. 2013;**43**:827-831. DOI: 10.1016/j.indcrop.2013.04.004

[88] Hossain MA, Al-Mijizy ZH, Al-Raqmi KAS, Weli AM, Al-Riyami Q. Study of total phenol, flavonoids contents and phytochemical screening of various leaves crude extracts of locally grown *Thymus vulgaris*. Asian Pacific Journal of Tropical Biomedicine. 2013;**3**(9):705-710

[89] Opara EI, Chohan M. Culinary herbs and spices: Their bioactive properties, the contribution of polyphenols and the challenges in deducing their true health benefits. International Journal of Molecular Sciences. 2014;**15**(10):19183-19202. DOI: 10.3390/ijms151019183

[90] El-Nekeety AA, Mohamed SR, Hathout AS, Hassan NS, Aly SE, Abdel-Wahhab MA. Antioxidant properties of *Thymus vulgaris* oil against aflatoxin-induce oxidative stress in male rats. Toxicon. 2011;**57**:984-991. DOI: 10.1016/j.toxicon.2011.03.021

[91] Karami-Osboo R, Khodaverdi M, Ali-Akbari F. Antibacterial effect of effective compounds of *Satureja hortensis* and *Thymus vulgaris* essential oils against *Erwinia amylovora*. Journal of Agricultural Science and Technology. 2010;**12**:35-45

[92] Bosquez-Molina E, Jesus ER, Bautista-Banos S, Verde-Calvo JR, Morales-Lopez J. Inhibitory effect of essential oils against *Colletotrichum gloeosporioides* and *Rhizopus stolonifer* in stored papaya fruits and their possible application in coatings. Postharvest Biology and Technology. 2010;**57**:132-137. DOI: 10.1016/j.postharvbio.2010.03.008

[93] Arras G, Usai M. Fungitoxic activity of 12 essential oils against four post-harvest citrus pathogens, chemical analysis of *Thymus capitatus* oil and its effect in sub-atmospheric pressure conditions. Journal of Food Protection. 2001;**64**:1025-1029. DOI: 10.4315/0362-028X-64.7.1025

[94] Satya P, Murray BL, McFeeters RL, Setzer WN. Essential oil characterization of *Thymus vulgaris* from various geographical locations. Foods. 2016;**5**(4):70. DOI: 10.3390/foods5040070

[95] Khumaloa KN, Tinyanea P, Soundya P, Romanazzi G, Glowacz M, Sivakumar D. Effect of thyme oil vapour exposure on the brown rot infection, phenylalanine ammonia-lyase (PAL) activity, phenolic content and antioxidant activity in red and yellow skin peach cultivars. Scientia Horticulturae. 2017;**214**:195-199. DOI: 10.1016/j.scienta.2016.11.044

[96] Soković M, Glamočlija J, Marin PD, Brkić D, van Griensven LJLD. Antibacterial effects of the essential oils of commonly consumed medicinal herbs using an in vitro model. Molecules. 2010;**15**:7532-7546. DOI: 10.3390/molecules15117532

[97] Upton R, Dayu RH. Stinging nettles leaf (*Urtica dioica* L.): Extraordinary vegetable medicine. Journal of Herbal Medicine. 2013;**3**(1):9-38. DOI: 10.1016/j.hermed.2012.11.001

[98] Gül S, Demirci B, Baser KH, Akpulat HA, Aksu P. Chemical composition and in vitro cytotoxic, genotoxic effects of essential oil from *Urtica dioica* L. Bulletin of Environmental Contamination and Toxicology. 2012;**88**(5):666-671. DOI: 10.1007/s00128-012-0535-9

[99] Gülcin I, Küfrevioglu ÖI, Oktay M. Purification and characterization of polyphenol oxidase from nettle (*Uritca dioica* L.) and inhibitory effects of some chemicals on enzyme activity. Journal of Enzyme Inhibition and Medicinal Chemistry. 2005;**20**:297-302. DOI: 10.1080/1475636032000141890

[100] Turker AU, Usta C. Biological screening of some Turkish medicinal plant extracts for antimicrobial and toxicity activities. Natural Product Research. 2008;**22**:136-146. DOI: 10.1080/14786410701591663

[101] Modarresi-Chahardehi A, Ibrahim D, Fariza-Sulaiman S, Mousavi L. Screening antimicrobial activity of various extracts of *Urtica dioica*. Revista de Biologia Tropical. 2012;**60**(4):1567-1576

[102] Gülçina I, Küfrevioğlu ÖI, Oktay M, Büyükokuroğluc ME. Antioxidant, antimicrobial, antiulcer and analgesic activities of nettle (*Urtica dioica* L.). Journal of Ethnopharmacology. 2004;**90**:205-215. DOI: 10.1016/j.jep.2003.09.028

[103] Antolak H, Czyzowska A, Kregiel D. Antibacterial and antiadhesive activities of extracts from edible plants against soft drink spoilage by *Asaia* spp. Journal of Food Protection. 2017;**80**(1):25-34. DOI: 10.4315/0362-028X.JFP-16-13

[104] Veberic R, Jakopic J, Stampar F, Schmitzer V. European elderberry (*Sambucus nigra* L.) rich in sugars, organic acids, anthocyanins and selected polyphenols. Food Chemistry. 2009;**114**:511-515. DOI: 10.1016/j.foodchem.2008.09.080

[105] Dawidowicz AL, Wianowska D, Baraniak B. The antioxidant properties of alcoholic extracts from *Sambucus nigra* L. (antioxidant properties of extracts). LWT—Food Science and Technology. 2006;**39**:308-315. DOI: 10.1016/j.lwt.2005.01.005

[106] Bhattacharya S, Christensen KB, Olsen LCB, ChristenjournalLP, Grevsen K, Færgeman NJ, Kristiansen K, Young JF, Oksbjerg N. Bioactive components from flowers of *Sambucus nigra* L. increase glucose uptake in primary porcine myotube cultures and reduce fat accumulation in *Caenorhabditis elegans*. Journal of Agricultural and Food Chemistry. 2013;**61**:11033-11040. DOI: 10.1021/jf402838a

[107] Sidor A, Gramza-MichaŁowska A. Advanced research on the antioxidant and health benefit of elderberry (*Sambucus nigra*) in food—a review. Journal of Functional Foods. 2014;**18**:941-958. DOI: 10.1016/j.jff.2014.07.012

[108] Mohammadsadeghi S, Malekpour A, Zahedi S, Eskanndari F. The antimicrobial activity of elderberry (*Sambucus nigra* L.) extract against gram positive bacteria, gram negative bacteria and yeast. Research Journal of Applied Sciences. 2013;**8**:240-243. DOI: 10.3923/rjasci.2013.240.243

[109] Hearst C, McCollum G, Nelson D, Ballard LM, Millar BC, Goldsmith CE, Roone PJ, Moore JE, Rao JR. Antibacterial activity of elder (*Sambucus nigra* L.) flower or berry against hospital pathogens. Journal of Medicinal Plants Research. 2010;**4**:1805-1809. DOI: 10.5897/JMPR10.147

[110] Unlu M, Ergene E, Unlu GV, Zeytinoglu HS, Vural N. Composition, antimicrobial activity and in vitro cytotoxicity of essential oil from *Cinnamomum zeylanicum* Blume (*Lauraceae*). Food and Chemical Toxicology. 2010;**48**(11):3274-3280. DOI: 10.1016/j.fct.2010.09.001

[111] Mallavarapu GR, Rajeswara Rao BR. Chemical constituents and uses of *Cinnamomum zeylanicum* Blume. In: Jitrovetz L, Dung NX, Varshney VK, editors. Aromatic Plants from Asia: Their Chemistry and Application in Food and Therapy. Har Krishan Bhalla & Sons, Prem Nagar, India; 2007. pp. 49-75

[112] Gupta C, Garg AP, Uniyal RC, Kumari A. Comparative analysis of the antimicrobial activity of cinnamon oil and cinnamon extract on some food-borne microbes. African Journal of Microbiology Research. 2008;**9**:247-251

[113] Jayaprakasha GK, Ohnishi-Kameyama M, Ono H, Yoshida M, Jaganmohan Rao L. Phenolic constituents in the fruits of *Cinnamomum zeylanicum* and their antioxidant activity. Journal of Agricultural and Food Chemistry. 2006;**54**(5):1672-1679. DOI: 10.1021/jf052736r

[114] Ranasinghe P, Pigera S, Premakumara GAS, Galappaththy P, Constantine GR, Katulanda P. Medicinal properties of 'true' cinnamon (*Cinnamomum zeylanicum*): A systematic review. BMC Complementary and Alternative Medicine. 2013;**13**:275. DOI: 10.1186/1472-6882-13-275

[115] Bayoub K, Baibai T, Mountassif D, Retmane A, Soukri A. Antibacterial activities of the crude ethanol extracts of medicinal plants against *Listeria monocytogenes* and some other pathogenic strains. African Journal of Biotechnology. 2010;**9**:4251-4258

[116] Elumalai S, Kesavan R, Ramganesh S, Prakasam V, Murugasen R. Comparative study on anti-microbial activities of bark oil extract from *Cinnamomum cassia* and *Cinnamomum zeylanicum*. Biosciences Biotechnology Research Asia. 2010;**7**:251-258

[117] Hosseininejad Z, Moghadam SD, Ebrahimi F, Abdollahi M, Zahedi MJ, Nazari M, Hayatbakhsh M, Adeli S, Sharififar F. In vitro screening of selected Iranian medicinal plants against *Helicobacter pylori*. International Journal of Green Pharmacy. 2011;**5**:282-285. DOI: 10.4103/0973-8258.94348

[118] Khan R, Islam B, Akram M, Shakil S, Ahmad A, Ali SM, Siddiqui M, Khan AU. Antimicrobial activity of five herbal extracts against multi drug resistant (MDR) strains of bacteria and fungus of clinical origin. Molecules. 2009;**14**:586-597. DOI: 10.3390/molecules14020586

[119] Tekwu EM, Askun T, Kuete V, Nkengfack AE, Nyasse B, Etoa FX, Beng VP. Antibacterial activity of selected Cameroonian dietary spices ethno-medically used against strains of mycobacterium tuberculosis. Journal of Ethnopharmacology. 2012;**142**:374-382. DOI: 10.1016/j.jep.2012.05.003

[120] Quale JM, Landman D, Zaman MM, Burney S, Sathe SS. In vitro activity of *Cinnamomum zeylanicum* against azole resistant and sensitive candida species and a pilot study of cinnamon for oral candidiasis. The American Journal of Chinese Medicine. 1996;**24**:103-109. DOI: 10.1142/S0192415X96000153

[121] Bhowmik D, Kumar KPS, Yadav A, Srivastava S, Paswan S, Dutta AS. Recent trends in Indian traditional herbs *Syzygium aromaticum* and its health benefits. Journal of Pharmacognosy and Phytochemistry. 2012;**1**:13-22

[122] Cortés-Rojas DF, Fernandes de Souza CR, Oliveira WP. Clove (*Syzygium aromaticum*): A precious spice. Asian Pacific Journal of Tropical Biomedicine. 2014;**4**(2):90-96. DOI: 10.1016/S2221-1691(14)60215-X

[123] Chaieb K, Hajlaoui H, Zmantar T, Kahla-Nakbi AB, Rouabhia M, Mahdouani K, Bakhrouf A. The chemical composition and biological activity of clove essential oil, *Eugenia caryophyllata* (*Syzigium aromaticum* L. *Myrtaceae*): A short review. Phytotherapy Research. 2007;**21**(6):501-506. DOI: 10.1002/ptr.2124

[124] Sofia K, Prasad R, Vijay VK, Srivastava AK. Evaluation of antibacterial activity of Indian spices against common foodborne pathogens. International Journal of Food Science and Technology. 2007;**42**(8):910-915. DOI: 10.1111/j.1365-2621.2006.01308.x

[125] Hill LE, Gomes C, Taylor TM. Characterization of beta-cyclodextrin inclusion complexes containing essential oils (trans-cinnamaldehyde, eugenol, cinnamon bark, and clove bud extracts) for antimicrobial delivery applications. LWT—Food Science and Technology. 2013;**51**(1):86-93. DOI: 10.1016/j.lwt.2012.11.011

[126] Devi KP, Nisha SA, Sakthivel R, Pandian SK. Eugenol (an essential oil of clove) acts as an antibacterial agent against *Salmonella typhi* by disrupting the cellular membrane. Journal of Ethnopharmacology. 2010;**130**(1):107-115. DOI: 10.1016/j.jep.2010.04.025

[127] Rana IS, Rana AS, Rajak RC. Evaluation of antifungal activity in essential oil of the *Syzygium aromaticum* (L.) by extraction, purification and analysis of its main component eugenol. Brazilian Journal of Microbiology. 2011;**42**(4):1269-1277. DOI: 10.1590/S1517-83822011000400004

[128] Park MJ, Gwak KS, Yang I, Choi WS, Jo HJ, Chang JW, Jeung EB, Choi IG. Antifungal activities of the essential oils in *Syzygium aromaticum* (L.) Merr. Et Perry and *Leptospermum petersonii* Bailey and their constituents against various dermatophytes. Journal of Microbiology. 2007;**45**(5):460-465

[129] Fu Y, Chen J, Li YJ, Zheng YF, Li P. Antioxidant and anti-inflammatory activities of six flavonoids separated from licorice. Food Chemistry. 2013;**141**(2):1063-1071. DOI: 10.1016/j.foodchem.2013.03.089

[130] Zhang Q-Y, Ye M. Chemical analysis of the Chinese herbal medicine Gan-Cao (licorice). Journal of Chromatography A. 2009;**1216**:1954-1969. DOI: 10.1016/j.chroma.2008.07.072

[131] Park SJ, Song HY, Youn HS. Suppression of the TRIF-dependent signaling pathway of toll-like receptors by isoliquiritigenin in RAW264.7 macrophages. Molecules and Cells. 2009;**28**:365-368. DOI: 10.1007/s10059-009-0130-z

[132] Fu Y, Hsieh TC, Guo J, Kunicki J, Lee MYWT, Darzynikiewicz Z, Wu JM. Licochalcone-A, a novel flavonoid isolated from licorice root (*Glycyrrhiza glabra*), causes G2 and late-G1 arrests in androgen-independent PC-3 prostate cancer cells. Biochemical and Biophysical Research Communications. 2004;**322**:263-270. DOI: 10.1016/j.bbrc.2004.07.094

[133] Asl MN, Hosseinzadeh H. Review of pharmacological effects of *Glycyrrhiza* sp. and its bioactive compounds. Phytotherapy Research. 2008;**22**(6):709-724. DOI: 10.1002/ptr.2362

[134] Friis-Møller A, Chen M, Fuursted K, Christensen SB, Kharazmi A. In vitro antimycobacterial and antilegionella activity of licochalcone A from Chinese licorice roots. Planta Medica. 2002;**68**:416-419. DOI: 10.1055/s-2002-32087

[135] Slavin LJ, Lloyd B. Health benefits of fruits and vegetables. Advances in Nutrition. 2012;**3**(4):506-516. DOI: 10.3945/an.112.002154

[136] Vattem DA, Ghaedian R, Shetty K. Enhancing health benefits of berries through phenolic antioxidant enrichment: Focus on cranberry. Asia Pacific Journal of Clinical Nutrition. 2005;**14**(2):120-130

[137] Costa AGV, Garcia-Diaz DF, Jimenez P, Silva PI. Bioactive compounds and health benefits of exotic tropical red–black berries. Journal of Functional Foods. 2013;**5**:539-549. DOI: 10.1016 /j.jff.2013.01.029

[138] González-Aguilar G, Robles-Sánchez RM, Martínez-Téllez MA, Olivas GI, Alvarez-Parrilla E, de la Rosa LA. Bioactive compounds in fruits: Health benefits and effect of storage conditions. Stewart Postharvest Review. 2008;**4**(3):1-10. DOI: 10.2212/spr.2008.3.8

[139] Manganaris GA, Goulas V, Vicente AR, Terry LA. Berry antioxidants: Small fruits providing large benefits. Journal of the Science of Food and Agriculture. 2014;**94**(5):825-833. DOI: 10.1002/jsfa.6432

[140] Szajdek A, Borowska EJ. Bioactive compounds and health-promoting properties of berry fruits: A review. Plant Foods for Human Nutrition. 2008;**63**:147-156. DOI: 10.1007/s11130-008-0097-5

[141] Antolak H, Kregiel D, Czyzowska A. Adhesion of *Asaia bogorensis* to glass and polystyrene in the presence of cranberry juice. Journal of Food Protection. 2015;**78**(6):1186-1190. DOI: 10.4315/0362-028X.JFP-14-440

[142] Blumberg JB, Camesano TA, Cassidy A, Kris-Etherton P, Howell A, Manach C, Ostertag LM, Sies H, Skulas-Ray A, Vita JA. Cranberries and their bioactive constituents in human health. Advances in Nutrition. 2013;**4**(6):618-632. DOI: 10.3945/an.113.004473a

[143] Palombo EA. Traditional medicinal plant extracts and natural products with activity against oral bacteria: Potential application in the prevention and treatment of oral diseases. Evidence-Based Complementary and Alternative Medicine. 2011:680354. DOI: 10.1093/ecam/nep067

[144] LaPlante KL, Sarkisian SA, Woodmansee S, Rowley DC, Seeram NP. Effects of cranberry extracts on growth and biofilm production of *Escherichia coli* and *Staphylococcus* species. Phytotherapy Research. 2012;**26**(9):1371-1374. DOI: 10.1002/ptr.4592

[145] Harich M, Maherani B, Salmieri S, Lacroix M. Antibacterial activity of cranberry juice concentrate on freshness and sensory quality of ready to eat (RTE) foods. Food Control. 2017;**75**:134-144. DOI: 10.1016/j.foodcont.2016.11.038

[146] Ermis E, Hertel C, Schneider C, Carle R, Stintzing F, Schmidt H. Characterization of in vitro antifungal activities of small and American cranberry (*Vaccinium oxycoccos* L. and *V. macrocarpon* Aiton) and lingonberry (*Vaccinium vitis-idaea* L.) concentrates in sugar reduced fruit spreads. International Journal of Food Microbiology. 2015;**204**:111-117. DOI: 10.1016/j.ijfoodmicro

[147] Moze S, Polak T, Gasperlin L, Koron D, Vanzo A, Poklar Ulrih N, Abram V. Phenolics in Slovenian bilberries (*Vaccinium myrtillus* L.) and blueberries (*Vaccinium corymbosum* L.). Journal of Agricultural and Food Chemistry. 2011;**59**(13):6998-7004. DOI: 10.1021/jf200765n

[148] Szakiel A, Pączkowski C, Huttunen S. Triterpenoid content of berries and leaves of bilberry *Vaccinium myrtillus* from Finland and Poland. Journal of Agricultural and Food Chemistry. 2012;**60**(48):11839-11849. DOI: 10.1021/jf3046895

[149] Seeram NP. Berry fruits for cancer prevention: Current status and future prospects. Journal of Agricultural and Food Chemistry. 2008;**56**(3):630-635. DOI: 10.1021/jf072504n

[150] Subash S, Essa MM, Al-Adawi S, Memon MA, Manivasagam T, Akbar M. Neuroprotective effects of berry fruits on neurodegenerative diseases. Neural Regeneration Research. 2014;**9**(16):1557-1566. DOI: 10.4103/1673-5374.139483

[151] Mykkänen OT, Huotari A, Herzig KH, Dunlop TW, Mykkänen H, Kirjavainen PV. Wild blueberries (*Vaccinium myrtillus*) alleviate inflammation and hypertension associated with developing obesity in mice fed with a high-fat diet. PLoS One. 2014;**9**(12):e114790. DOI: 10.1371/journal.pone.0114790

[152] Burdulis D, Sarkinas A, Jasutienė I, Stackevicenė E, Nikolajevas L, Janulis V. Comparative study of anthocyanin composition, antimicrobial and antioxidant activity in bilberry (*Vaccinium myrtillus* L.) and blueberry (*Vaccinium corymbosum* L.) fruits. Acta Poloniae Pharmaceutica. 2009;**66**(4):399-408

[153] Nohynek LJ, Alakomi HL, Kähkönen MP, Heinonen M, Helander IM, Oksman-Caldentey KM, Puupponen-Pimiä RH. Berry phenolics: Antimicrobial properties and mechanisms of action against severe human pathogens. Nutrition and Cancer. 2006;**54**(1):18-32. DOI: 10.1207/s15327914nc5401_4

[154] Heinonen M. Antioxidant activity and antimicrobial effect of berry phenolics—a Finnish perspective. Molecular Nutrition and Food Research. 2007;**51**(6):684-691. DOI: 10.1002/mnfr.200700006

[155] Puupponen-Pimiä R, Nohynek L, Meier C, Kähkönen M, Heinonen M, Hopia A, Oksman-Caldentey KM. Antimicrobial properties of phenolic compounds from berries. Journal of Applied Microbiology. 2001;**90**(4):494-507

[156] Donno D, Beccaro GL, Mellano MG, Cerutti AK, Marconi V, Bounous G. Botanicals in *Ribes nigrum* bud-preparations: An analytical fingerprinting to evaluate the bioactive contribution to total phytocomplex. Pharmaceutical Biology. 2013;**51**(10):1282-1292. DOI: 10.3109/13880209.2013.786101

[157] Sójka M, Guyot S, KoŁodziejczyk K, Król B, Baron A. Composition and properties of purified phenolics preparations obtained from an extract of industrial black currant (*Ribes nigrum* L.) pomace. Journal of Horticultural Science and Biotechnology. 2009;**84**(6):100-106 [ISAFRUIT Special Issue]

[158] Khoo GK, Clausen MR, Pedersen HL, Larsen E. Bioactivity and chemical composition of black currant (*Ribes nigrum*) cultivars with and without pesticide treatment. Food Chemistry. 2012;**132**(1):1214-1220. DOI: 10.1016/j.foodchem.2011.11.087

[159] Wu QK, Koponen JM, Mykkanen HM, Törrönen AR. Berry phenolic extracts modulate the expression of p21(WAF1) and bax but not Bcl-2 in HT-29 colon cancer cells. Journal of Agricultural and Food Chemistry. 2007;**55**:1156-1163. DOI: 10.1021/jf062320t

[160] Boivin D, Blanchett M,journalrette S, Moghrabi A, Beliveau R. Inhibition of cancer cell proliferation and suppression of TNF-induced activation of NFkappaB by edible berry juice. Anticancer Research. 2007;**27**(2):937-948

[161] Galgóczy L, Hevér T, Orosz L, Krisch J, Vágvölgyi C, Tölgyesi M, Papp T. Growth inhibition effect of fruit juices and pomace extracts on the enteric pathogens *Campylobacter jejuni* and *Salmonella* ser. *Typhimurium*. The Internet Journal of Microbiology. 2008;**7**(1): 47570393. Available from: http://ispub.com/IJMB/7/1/7317 [Accessed: 25 February]

[162] Krisch J, Galgóczy L, Tölgyesi M, Papp T, Vágvölgyi C. Effect of fruit juices and pomace extracts on the growth of gram-positive and gram-negative bacteria. Acta Biologica Szegediensis. 2008;**52**(2):267-270

[163] Ikuta K, Hashimoto K, Kaneko H, Mori S, Ohashi K, Suzutani T. Anti-viral and anti-bacterial activities of an extract of black currants (*Ribes nigrum* L.). Microbiology and Immunology. 2012;**56**(12):805-809. DOI: 10.1111/j.1348-0421.2012.00510.x

[164] Krisch J, Ördögh L, Galgóczy L, Papp T, Vágvölgyi C. Anticandidal effect of berry juices and extracts from *Ribes* species. Central European Journal of Biology. 2009;**4**(1):86-89. DOI: 10.2478/s11535-008-0056-z

[165] Charlebois D, Byers PL, Finn CE, Thomas AL. Elderberry: Botany, horticulture, potential. In: Janick J, editor. Horticultural Reviews. Wiley-Blackwell, New Jersey, USA 2010. pp. 213-280. DOI: 10.1002/9780470543672.ch4

[166] Roschek B Jr, Fink RC, McMichael MD, Li D, Alberte RS. Elderberry flavonoids bind to and prevent H1N1 infection in vitro. Phytochemistry. 2009;**70**(10):1255-1261. DOI: 10.1016/j.phytochem.2009.06.003

[167] Krawitz C, Mraheil MA, Stein M, Imirzalioglu C, Domann E, Pleschka S, Hain T. Inhibitory activity of a standardized elderberry liquid extract against clinically-relevant human respiratory bacterial pathogens and influenza A and B viruses. BMC Complementary and Alternative Medicine. 2011;**11**:16. DOI: 10.1186/1472-6882-11-16

[168] Arjoon AV, Saylor CV, May M. In vitro efficacy of antimicrobial extracts against the atypical ruminant pathogen *Mycoplasma mycoides* subsp. *capri*. BMC Complementary and Alternative Medicine. 2012;**12**:169. DOI: 10.1186/1472-6882-12-169

[169] Chatterjee A, Yasmin T, Bagchi D, Stohs SJ. Inhibition of *Helicobacter pylori* in vitro by various berry extracts, with enhanced susceptibility to clarithromycin. Molecular and Cellular Biochemistry. 2004;**265**(1-2):19-26

[170] Rop O, Mlcek J, Kramarova D, Jurikova T. Selected cultivars of cornelian cherry (*Cornus mas* L.) as a new food source for human nutrition. African Journal of Biotechnology. 2010;**8**:1205-1210

[171] Milenković-Andjelković AS, Andjelković MZ, Radovanović AN, Radovanović BC, Nikolić V. Phenol composition, DPPH radical scavenging and antimicrobial activity of cornelian cherry (*Cornus mas*) fruit and leaf extracts. Hemijska Industrija. 2015;**69**(4):331-337. DOI: 10.2298/HEMIND140216046M

[172] Tural S, Koca I. Physico-chemical and antioxidant properties of cornelian cherry fruits (*Cornus mas* L.) grown in Turkey. Scientia Horticulturae. 2008;**116**:362-366. DOI: 10.1016/j.scienta.2008.02.003

[173] Popović BM, Stajner D, Slavko K, Sandra B. Antioxidant capacity of cornelian cherry (*Cornus mas* L.)—comparison between permanganate reducing antioxidant capacity and other antioxidant methods. Food Chemistry. 2012;**134**(2):734-741. DOI: 10.1016/j.foodchem

[174] Francik R, Kryczyk J, Krośniak M, Berköz M, Sanocka I, Francik S. The neuroprotective effect of *Cornus mas* on brain tissue of Wistar rats. The Scientific World Journal. 2014;**2014**:847368. DOI: 10.1155/2014/847368

[175] Kyriakopoulos AM, Dinda B. *Cornus mas* (*Linnaeus*) Novel devised medicinal preparations: Bactericidal effect against *Staphylococcus aureus* and *Pseudomonas aeruginosa*. Molecules. 2015;**20**(6):11202-11218. DOI: 10.3390/molecules200611202

[176] Gruenwald I. Novel botanical ingredients for beverages. Clinics in Dermatology. 2009;**27**(2):210-216. DOI: 10.1016/j.clindermatol.2008.11.003

Chapter 4

Natural Antimicrobials, their Sources and Food Safety

Muhammad Sajid Arshad and Syeda Ayesha Batool

Additional information is available at the end of the chapter

http://dx.doi.org/10.5772/intechopen.70197

Abstract

With consumer awareness about food safety and quality, there is a high demand for the preservative (synthetic)-free foods and use of natural products as preservatives. Natural antimicrobials from different sources are used to preserve food from spoilage and pathogenic microorganisms. Plants (herbs and spices, fruits and vegetables, seeds and leaves) are the main source of antimicrobials and contain many essential oils that have preservation effect against different microorganisms. Mainly, herb and spices contain many essential oils and the examples include rosemary, sage, basil, oregano, thyme, cardamom, and clove. These essential oils are very effective against many pathogenic and spoilage microorganisms like *Salmonella, Escherichia coli, Listeria monocytogenes, Campylobacter* spp., and *Staphylococcus aureus* and help to increase their quality and shelf stability. These antimicrobial compounds are also used in combination with edible food coatings and inhibit the ability of microorganisms to grow on the surface of food and food products.

Keywords: antimicrobial, essential oils, antimicrobial edible coatings

1. Introduction

Today, food safety is everybody's concern and it is very hard to find anyone who has not encountered an unpleasant moment of food-borne illness at least once in the past year. According to the report of WHO in 2005, there were about 1.8 million deaths caused by diarrhea (food-borne illness), and these diseases were due to the use of contaminated food and water [1]. The main cause of food-borne illnesses is the use of food contaminated by microbial pathogens, toxins, or radioactive components. When certain bacteria or pathogens contaminate food, they can cause food-borne illness or sometimes called "food poisoning." Food-borne illnesses are mild but sometimes they can even be deadly [2].

Food-borne pathogens (*Clostridium botulinum, Staphylococcus aureus, Campylobacter jejuni, Bacillus cereus, Listeria monocytogenes, Cryptosporidium, Escherichia coli* O157:H7, etc.) are the main concern regarding the safety of food [3]. Food can contain microbiological pathogens that cause infections or intoxications, or chemical agents that cause acute or chronic intoxications. With special reference to meat and meat products, *Salmonella, E. coli, L. monocytogenes,* and *Campylobacter* are the main pathogenic organisms [4, 5].

There is an increase in the consumption of fresh food with the consumer demand for the ready-to-eat food and the desire to lead a healthy lifestyle. The challenges associated with the consumption of fresh food are short storage life and its association with food-borne diseases. To avoid these challenges, there is a commercial pressure of using chemical preservatives that prevent the growth of food spoilage agents, but the increase in the use of these chemical preservatives is negatively perceived by the consumer [6].

2. Antimicrobial agents and food safety

Traditional food preservation methods are less efficient in reducing the growth of food-borne pathogens in food products, and the ever-increasing demand for chemical-free food has paved the way for antimicrobials to be used in food industry [7]. The use of antimicrobials is a new technology by the food industry to increase the shelf life of food and overcome the issues of food quality and safety. These antimicrobials could be of natural or synthetic type, but natural antimicrobials are gaining much importance than synthetic ones. Even though synthetic preservatives are approved by government agencies for human use, many of these preservatives still threaten our health. Thus, researchers give more importance toward the potential of natural products for their antimicrobial activities [8–10].

3. Natural antimicrobial agents

Chemical compounds having pharmacological and biological activity and produced by living organisms are called natural products. Living organisms produce primary and secondary metabolites [11–13]. Primary metabolites are the products that have essential function in the organism, while secondary metabolites could simply be waste products or could have some important function in their producers. Secondary metabolites can be used as drugs against diseases such as cancer, inflammation (swelling), and so on and also have antimicrobial activity [1, 14]. Secondary metabolites possessing antimicrobial activity are called the natural antimicrobials and could be extracted from different sources like plants (fruits, vegetables, seeds, herb, and spices), animals (eggs, milk, and tissues), and microorganisms (fungi and bacteria) [15–17]. With special reference to plants, secondary metabolites are found to be healthy ingredients that work as antimicrobials or disease-controlling agents [4]. Owing to the potential of antimicrobials against pathogenic and spoilage microorganisms, these secondary metabolites gain much importance for the application in food products [18–20]. They contain the properties of antimicrobials and antioxidants at the same time and so are considered as a better option for food preservation as compared to synthetic preservatives [21].

Several researches have been conducted to find out the antimicrobial potential of natural products, especially the plant sources like fruits, vegetables, herbs, and spices because they are enriched with compounds having antimicrobial activity. Nowadays, there are more than 1350 plants with antimicrobial activities and more than 30,000 antimicrobial components have been extracted from plants [22]. However, many studies have also been conducted on antimicrobial potential of microorganisms and animals. Food applications of antimicrobials have also been investigated.

Nowadays, plant extracts and essential oils (EOs) have gained much importance due to their flavoring as well as antimicrobial potential [23]. Research conducted on the antimicrobial activity of the extracts from different fruit peels like banana, apple, pomegranate, sweet lime, orange, mango, and papaya indicated that fruit peel extracts have mild inhibitory effect against pathogenic bacteria [24–29]. Plants secondary metabolites contain many antimicrobial agents, so they have a greater inhibitory effect against Gram-positive and Gram-negative bacteria [14, 27, 30–32]. The chemical composition, concentration, and structure of the antimicrobial component determine their efficacy. Antimicrobial components of plant origin include flavonoids, thiosulfinates, glucosinolates, phenolics, organic acids, flavonoids, and saponins [31, 33, 34]. However, the main compounds with antimicrobial activity are phenols which include terpenes, aliphatic alcohols, aldehydes, ketones, acids, and isoflavonoids [35–38].

Antimicrobial components in plant materials are commonly found in herbs and spices (rosemary, sage, basil, oregano, thyme, cardamom, and clove), fruits and vegetables (guava, pepper, cabbage, garlic, and onion, citrus), seeds and leaves (grape seeds, fennel, nutmeg, parsley, and olive leaves) [39–42].

In this chapter, we discuss the role of antimicrobials from different sources with special reference to meat and meat products. Consumption of meat is important for the growth, development, and maintenance of health in human beings. Meat is an animal origin food and is a rich source of proteins, vitamins, minerals, and so on which is why the safety of meat and meat products is of much importance [43, 44]. Proteins of meat are of much importance with a high amount of essential amino acids being available and of biological value. Meat and meat products are at a high risk of microbial spoilage and also cause losses to economy [45]. Although food industry has developed several new techniques for hygienic slaughtering and production of meat products, a major concern related to meat consumption is the presence of pathogenic microorganisms that cause food-borne diseases, for which raw meat provides an ideal substrate [46, 47]. *Salmonella* spp., *Campylobacter* spp., *L. monocytogenes*, *E. coli*, and *S. aureus* are the most common meat spoilage agents that cause food-borne diseases worldwide [48]. Synthetic preservatives are used to overcome this problem, but their overuse leads to multidrug-resistant phenomenon in bacteria. Moreover, meat industry is facing a new trend of developing all natural food products, where there is no place for synthetic preservatives that could be the causative of food sensitivities, toxicities, and allergies [49–51].

Essential oils, as plant extracts possessing antimicrobial agents and also antioxidative and flavoring properties, can be considered as healthy ingredients to be used in meat and meat products. If essential oils are used in meat products, they can reduce the chances of food-borne diseases and can retard the oxidation of lipids in meat [52–54].

4. Antimicrobials from plant sources

4.1. Herbs and spices

Herbs and spices have long been used by human beings for different reasons like food additives, flavorings, and preservatives. They are considered the most commonly used natural antimicrobials against different pathogens. The antimicrobial activity of herbs and spices depends on the type of essential oil present in it, food type in which it has to be used, and the type of microorganism [11, 55–57].

The efficiency of essential oils from herbs and spices depends upon their chemical structure, in particular to the presence of hydrophilic functional groups such as hydroxyl groups [58]. Essential oils from clove, oregano, rosemary, thyme, sage, and vanillin are the most effective containing the phenolic groups [58]. They possess inhibitory activity against Gram-positive than Gram-negative bacteria [59, 60]. Essential oils have high vapor pressure and are able to reach pathogenic microorganism through gas or liquid phases. Many investigations have proved the antimicrobial efficiency of essential oils against several pathogenic and spoilage microflorae. However, the efficiency of essential oils depends upon the pH, storage temperature, and concentration of oxygen [61].

Some of the antimicrobial compounds that are present in spices and herbs are eugenol, thymol, thymol and carvacrol, vanillin, allicin, cinnamic aldehyde, and allyl isothiocyanate that are, respectively, present in cloves, thyme, oregano, vanilla, garlic, cinnamon, and mustard [26].

Essential oils possess antimicrobial activities against several pathogenic microorganisms present in meat, including both Gram-positive and Gram-negative bacteria [62]. Many studies have been conducted to analyze the effects of essential oils extracted from sources such as oregano, rosemary, thyme, basil, garlic, and clove, when used alone or in combination with other essential oils [4, 63].

Essential oils extracted from herbs and spices were found to be effective against several pathogenic microorganisms. Studies showed the antimicrobial activities of 14 essential oils (clove, oregano, rosemary, pepper, nutmeg, liquorice, turmeric, aniseed, cassia bark, fennel, prickly ash, round cardamom, dahurian angelipca root, and angelica) against four meat spoilage and pathogenic bacteria (*L. monocytogenes, E. coli, Pseudomonas fluorescens* and *Lactobacillus sake*), and the results showed that extracts of clove, rosemary, and cassia bark contained strong antimicrobial activity against these bacteria but a combination of rosemary and liquorice extracts was the best inhibitor against all four types of bacteria. Antimicrobials from herbs and spices are widely used by the industry, and government agencies have approved them to be safe [64, 65]. *Pseudomonas* bacteria are responsible for the unacceptability of meat sausages. The use of thymol extracted from thyme and oregano as an antimicrobial inhibits the growth of *Pseudomonas* in sausages [53]. Researches show that Marjoram, mustard, cinnamon, lemon grass, and rosemary extracts have inhibitory effects against *E. coli O157:H7, S. typhi,* and *Listeria* [66, 67].

The oregano essential oils have antibacterial activities against *E. coli, S. aureus, B. subtilis,* and *Saccharomyces cerevisiae*. The main component in the essential oil of oregano is carvacrol (80.5%). Some other studies show that *S. typhimurium* is more sensitive to oregano essential oils than *S. enteritidis* [68].

Sodium nitrite has been used as a preservative in meat and meat products, but researches showed that if it is used in combination with oregano essential oils, it will slow down the growth of bacteria more efficiently than sodium nitrite alone [14]. The amount of EOs used in meat and meat products should be higher than the dose used in in vitro conditions because of the interaction with components of meat. Antimicrobial essential oils can be used directly or as polyethylene oxide (PEO)-based antimicrobial packaging [69].

Another research showed that the addition of oregano essential oils at a concentration of 0.7% will provide antimicrobial activity in minced sheep meat against *S. enteritidis*. In vitro tests detected the antibacterial activity of oregano oil against *S. enteritidis* in foods such as traditional salted fish and cod fish [70].

Mustard and horseradish essential oils also have antimicrobial activities against Gram-negative bacteria. Major antimicrobial agent in both is allyl-isothiocyanate [71–73].

When applying the antimicrobials in meat or meat products, depending upon the properties and type of pathogen, some EOs are more effective than others. Eugenol, coriander, clove, oregano, and thyme oils were found to be effective at levels of 5–20 μl/g in inhibiting *L. monocytogenes* in meat products, while mustard, mint, and sage oils were less effective or ineffective [74].

Rosmarinus officinalis L. commonly called rosemary is cultivated in southern Europe and is used as a flavoring agent due to its better flavor, high antioxidant, and antibacterial capacity [74, 75]. Carnosic acid and carnosol are the major antimicrobial components of rosemary and are effective against both Gram-negative and Gram-positive bacteria. In meat and meat products, rosemary oil has high antibacterial activity against *L. monocytogenes* [76].

Thyme essential oils have high antimicrobial activity owing to the presence of different compounds. The most prominent of all identified compounds of thyme essential oils were thymol (50%), followed by p-cymene (24%), linalool (4.6%), γ-terpinene (4.1%), and 1,8-cineole (4.3%). Thyme oils are effective against *L. monocytogenes* at a dose level of 5–20 μl/g. When added at a dose level of 0.3–0.9%, they are very effective against *E. coli* in beef. In vitro antimicrobial activity of thyme essential oil has been tested against *E. coli* at a temperature higher than that of refrigeration [77, 78].

Extensive research has been conducted to analyze the efficiency of essential oils against *Salmonella*, and results showed that oils extracted from thyme and oregano reduce the growth of *Salmonella* up to many folds of colony-forming unit (CFU) levels, while cinnamon oils at a rate of 7000 mg/kg of meat have strong antibacterial activity against *Salmonella* [62, 77].

Research has been conducted to find out the antimicrobial activity of clove oil against *L. monocytogenes* in minced mutton. Thymol essential oil from thyme at a concentration of

250–750 mg is used in fresh minced beef in combination with modified atmosphere packaging against different microorganisms and also increases the shelf life of beef [26, 29].

Sage essential oil is used at a concentration of 0.3% in minced beef in combination with soy protein. Rosemary or Chinese mahogany (500, 1000, and 1500 ppm) is used to increase fresh chicken sausage [14, 69, 79].

4.1.1. Safety aspect of essential oils

Antimicrobial agents, though very effective against microbial population and able to extend the quality and shelf life of meat and meat products, should be added with care because they can cause some side effects. Many essential oils like thymol and eugenol can cause mucous membrane irritation, if used in higher concentrations. In vitro studies of various essential oils like carvacrol, carvone, thymol, and so on show a mild to moderate toxic effects [30]. Some essential oils can cause allergy or some can have photoactive molecules which can cause phototoxic reactions [80, 81].

4.2. Fruits and vegetables

Many fruits and vegetables are nowadays well known to have antimicrobial effect against different pathogenic and spoilage microbes due to their contents of phenolic and organic acids. Fruit peels that are mostly discarded also contain antimicrobial compounds [30, 82].

Research showed that the antimicrobial activity of orange peel and capsicum was due to the presence of phenolic compound (coumaric acid) [83]. In minced beef, the extracts of capsicum annum have inhibitory effect against *S. typhimurium* and *Pseudomonas*. The minimum dose level of capsicum extract was 1.5 ml/100 g of minced beef to inhibit the growth of *S. typhimurium*, while a dose of 3 ml/100 g was required for a bactericidal effect against *P. aeruginosa* [83].

Pomegranate extract reduces the growth of *E. coli*. The peel of pomegranate contains different phenols and flavonoids that have great antimicrobial activity against Gram-positive bacteria. Peel extracts have inhibitory effects against *S. aureus* and *B. cereus* at a concentration of 0.01%. The addition of pomegranate peel extract to chicken meat products increases its shelf stability by 2–3 weeks during chilled storage and its extract is also effective in controlling oxidative rancidity in these chicken products [7, 62].

Citrus peel extract, lemon grass, and lime peel extracts were investigated for their antimicrobial activities in meat and meat products. The extracts showed high potential of antibacterial activity against *B. cereus*, *S. typhimurium*, and *S. aureus*. Hot water extract of lemon fruit peels, seeds, and juices displayed promising evidence of antibacterial activity against bacteria *E. coli*, *P. aeruginosa*, and *S. aureus* [84, 85].

Garlic is a potential inhibitor for food pathogens. Foods contaminated with pathogens pose a potential danger to the consumer's health. The use of garlic can increase the shelf life and decrease the possibilities of food poisoning and spoilage in processed foods. Garlic extract has antimicrobial activity due to the presence of an organic sulfur compound allicin, which acts as a growth inhibitor for both Gram-positive and Gram-negative bacteria including *E. coli, Salmonella, Streptococcus, Staphylococcus, Klebsiella, Proteus, and Helicobacter pylori.* Garlic aqueous extract has antibacterial properties against *S. aureus* present in hamburger. Freshly ground garlic in combination with lean camel meat at a concentration of 10, 15, and 25% was used to increase the shelf life of meat at different temperatures (rooms, incubators, refrigerators). After 4 days of storage at room temperature, 12 days of incubation, and 28 days of refrigeration, it was found that treatments with 15 and 25% garlic resulted in complete inhibition of microbial growth with no sign of any organoleptic spoilage of the meat [11].

The antimicrobial effect of onion extract on the fresh beef fillet meat was investigated. Beef fillet samples were cut into pieces and treated with 5, 10, 20, and 50% onion-water extract (v/v) and stored in refrigeration conditions at 4°C. Microbiological quality of the samples was investigated during storage for 9 days. Increasing concentrations of onion extract significantly affected *E. coli* and yeast-mold counts, but *Pseudomonas* spp., aerobic mesophilic bacteria, and total coli forms were not affected significantly for some concentrations and days.

Antimicrobial efficacy of curcumin, one of the active components of the *Curcuma longa* (turmeric) plant, was evaluated against food pathogens in a minced meat medium. *S. typhimurium, L. monocytogenes, E. coli* O157:H7, and *S. aureus* strains were used as food pathogens [86].

5. Antimicrobial edible coatings

Today, many fresh products are available commercially with best nutritional profile and low cost of production. Consumers also prefer consuming fresh meat and meat products, but a limit for the commercial availability of fresh meat is its low storage life because of high moisture contents that cause the growth of pathogenic and spoilage microorganisms [87].

To avoid this, the spoilage use of antimicrobials is one of the best ways to increase the shelf life of these perishable food products especially meat and meat products. The use of antimicrobial films and coatings dates back to twelfth century. The only difference between film and coating is its thickness. There are many ways of applying these antimicrobials on food products to enhance the natural appearance and safety of fresh meat and meat products like spray or spread of antimicrobials on meat [88, 89].

By the combination of different preservation techniques, researchers have been successful in achieving the objectives related to microbial quality storage life of perishable products. The addition of natural antimicrobials in combination with modified atmosphere packaging and refrigeration has proven to show the best results. Antimicrobials can also be added in coatings and films to be used in meat and meat products [88, 90].

The use of antimicrobials in edible films and coatings is an emergent technique that is helpful in enhancing the quality and safety aspect of food. This technique includes a control release of antimicrobial agents in effective concentration in the food product, when required.

The use of oregano essential oil (EO) as natural antimicrobial in combination with modified atmosphere packaging and refrigeration highly enhances the storage life of fresh beef or chicken during storage. Whey protein isolate coatings containing antimicrobial agents like oregano EO, 3-polylysine, or sodium lactate were used on fresh beef under refrigeration, which was evaluated against the progression of microflora like *Pseudomonas* bacteria [91]. By using 1.5% of oregano EO or 0.75% of 3-poly-lysine, the growth of *Pseudomonas* spp. was reduced and the development of lactic acid bacteria was completely inhibited. Both *Pseudomonas* spp. and total viable microorganisms were completely inhibited with 2% sodium lactate, even though the effect on LAB was less intense [89].

The effect of soy protein isolate films containing up to 5% of oregano and/or thyme EO was evaluated to be effective against coliform and *Pseudomonas* spp., but not significantly effective against total viable microorganisms, LAB, or *Staphylococcus* spp. in vacuum-packaged minced beef burgers for a 12-day period of cold storage at 4°C. Carvacol and cinnamon aldehydes, the main active compounds of oregano and cinnamon essential oils, were evaluated for their antimicrobial activity; they were incorporated in edible films based on apple puree containing 1.5 and 3% of carvacrol or cinnamaldehyde over chicken breast under refrigeration. These films inactivated the autochthonous spoilage microflora of chicken [89].

Whey protein isolates-based edible films were evaluated for antimicrobial activities with different essential oils. These films showed high effectiveness against *L. monocytogenes, E. coli* O157:H7, and *S. enterica Typhimurium*, when used in combination with 1% sorbic acid in meat sausages. Oregano containing carvacrol as antimicrobial agent and clove containing eugenol EOs were highly effective against *S. aureus, Salmonella*, and *L. innocula*. Coatings act as barrier against oxygen transfer leading to growth inhibition of aerobic bacteria. Chitosan has been used as an antimicrobial agent and also as a coating and wrapper in salami and film and coating combined with lauric arginate and nisin to reduce *L. monocytogenes* population in sliced turkey deli meat and also in seafood and fish [63, 64].

Similarly, milk protein coatings are used in beef in combination with oregano essential oils against *E. coli* and *Pseudomonas*. Chistosan coatings dissolved in lactic acid in combination with 1% acetic acid are used in roasted beef products against *L. monocytogenes*. Similarly, chitosan coatings in combination with oregano oil at a concentration of 0.7% are used against *Pseudomonas* spp. and *Brochothrix thermosphacta* [91]. Gelatin films are used in Turkey bologna in combination with nisaplin-based films (GNF) (0.025–0.5%; w/v nisin) against *L. monocytogenes* bacteria [89].

Chitosan coatings in different molecular weights and viscosities (14, 57, or 360 mPa) were used in Atlantic cod fish against psychotropic bacteria. Whey protein coatings were used in smoked fish in combination with Lactoperoxidase system (0–0.5%, w/v) against *L. monocytogenes*. Gelatin films were used in sardine pilchardus in combination with oregano extracts against Enterobacter bacteria. Alginate, carrageenan, pectin, gelatin, or starch coatings were also used in smoked salmons in combination with sodium lactate against a mixture of *L. monocytogenes* [92].

6. Conclusion

All the researches and studies conducted till now have proved that the use of synthetic preservatives to increase the shelf life of food and food products is in any way harmful for the human health, so there is a call for the use of natural products as preservatives to increase the quality and shelf stability of the food and food products. Natural antimicrobials contain all the qualities to be used as preservatives especially in meat and meat products, and plants are the main source of these antimicrobials.

Plant essential oils have great antimicrobial activity against Gram-positive and Gram-negative bacteria owing to the potential of phenolic compounds. Essential oils from herbs and spices like clove, oregano, rosemary, thyme, sage, and vanillin are the most effective against spoilage and pathogenic microorganisms like *L. monocytogenes*, *E. coli*, *P. fluorescens*, *L. sake*, *S. aureus*, and *B. subtilis*. Mustard and horseradish essential oils also have antimicrobial activities against Gram-negative bacteria. Major antimicrobial agent in both is allyl-isothiocyanate. Antimicrobial agents, though very effective as antimicrobial agents, should be used with care because they can cause side effects like irritation. Many fruits and vegetables also contain antimicrobial activity against pathogenic and spoilage microbes. Extracts of capsicum annum showed antimicrobial effects against *S. typhimurium* in minced beef; similarly, pomegranate extracts reduced the growth of *E. coli*. Citrus peel extract, lemon grass, and lime peel extracts showed great antimicrobial effect against *B. cereus*, *S. typhimurium*, and *S. aureus*. Garlic is a potential inhibitor for food pathogens. Garlic aqueous extract has antibacterial properties against *S. aureus* present in hamburger.

To increase the shelf life of meat and meat products, a new trend is the use of antimicrobial in edible films and coatings in combination with different packaging techniques. Oregano essential oils in combination with modified atmosphere packaging highly increase the shelf life of chicken and beef. Whey protein isolate coatings added with oregano essential oils in combination with refrigeration were very effective against *Pseudomonas* in beef and beef products. Whey protein isolate-based edible films were evaluated for antimicrobial activities with different essential oils and were very effective against *S. aureus*, *Salmonella*, and *L. innocula*. Antimicrobials can be sprayed upon meat and meat products or meat can be dipped into them. They are completely harmless to human health owing to the potential of all natural compounds, so there is an increasing market for the natural antimicrobials to be used as preservatives.

Author details

Muhammad Sajid Arshad* and Syeda Ayesha Batool

*Address all correspondence to: sajid_ft@yahoo.com

Institute of Home and Food Sciences, Government College University, Faisalabad, Pakistan

References

[1] Samelisa J. Managing microbial spoilage in the meat industry. In: Blackburn W, editor. Food Spoilage Microorganisms. Boca Raton, FL: CRC Press LLC; 2006

[2] Bingöl EB, Bostan K. Effect of sodium lactate on the microbiological quality and shelf life of sausages. Turkish Journal of Veterinary & Animal Sciences. 2007;**31**:33-39

[3] Demain AL. Microbial secondary metabolism: A new theoretical frontier for academia, a new opportunity for industry. Secondary Metabolites: Their Function and Evolution. 1992;**3**:21-23

[4] Atarés L, De Jesús P, Talens C, Chiralt A. Characterization of SPI-based edible films incorporated with cinnamon or ginger essential oils. Journal of Food Engineering. 2010;**99**:384-391

[5] Doyle MP, Erickson MC. Emerging microbiological food safety issues related to meat. Meat Science. 2006;**74**:98-112

[6] Tohidpour A, Sattari M, Omidbaigi R, Yadegar A, Nazemi J. Antimicrobial effect of essential oils from two medicinal plants agains methicillin-resistant *Staphylococcus aureus* (MRSA). Phytomedicine. 2010;**17**:142-145

[7] Jagtap NS, Khadabadi SS, Ghorpade DS, Banarase NB, Naphade SS. Antimicrobial and antifungal activity of *Centella asiatica* (L.) Urban, Umbeliferae. Research Journal of Pharmacology and Technology. 2009;**2**:328-330

[8] Elgayyar M, Draughon FA, Golden DA, Mount JR. Antimicrobial activity of essential oils from plants against selected pathogenic and saprophytic microorganisms. Journal of Food Protection. 2001;**64**:1019-1024

[9] Skandamis P, Tsigarida E, Nychas GJE. The effect of oregano essential oil on survival/death of *Salmonella typhimurium* in meat stored at 5°C under aerobic, VP/MAP conditions. Food Microbiology. 2002;**19**:97-103

[10] Dorman HJD, Deans SG. Antimicrobial agents from plants: Antibacterial activity of plant volatile oils. Journal of Applied Microbiology. 2009;**88**:308-316

[11] Astal ZE. The inhibitory action of aqueous garlic extract on the growth of certain pathogenic bacteria. European Food Research and Technology. 2004;**218**:460-464

[12] Economou T, Pournis N, Ntzimani A, Savvaidis IN. Nisin–EDTA treatments and modified atmosphere packaging to increase fresh chicken meat shelf-life. Food Chemistry. 2009;**114**:1470-1476

[13] Krug D, Zurek G. Discovering the hidden secondary metabolome of *Myxococcus xanthus*: A study of intraspecific diversity. Applied Environmental Microbiology. 2008;**74**:58-68

[14] Burt S. Essential oils: Their antibacterial properties and potential applications in foods—A review. International Journal of Food Microbiology. 2004;**3**:223-253

[15] Jałosńska M, Wilczak J. Influence of plant extracts on the microbiological shelf life of meat products. Polish Journal of Food and Nutrition Science. 2009;**59**:303-308

[16] Koehn FE, Carter GT. The evolving role of natural products in drug discovery. Nature Reviews Drug Discovery. 2005;**4**:206

[17] Medema MH, Blin K. AntiSMASH: Rapid identification, annotation and analysis of secondary metabolite biosynthesis gene clusters in bacterial and fungal genome sequences. Nucleic Acids Research. 2011;**39**:339-346

[18] Demain AL, Fang A. The natural functions of secondary metabolites. Advances in Biochemical Engineering/Biotechnology. 2000;**69**:1-39

[19] Fedorova ND, Moktali V, Medema MH. Bioinformatics approaches and software for detection of secondary metabolic gene clusters. Methods in Molecular Biology. 2012;**944**:23-45

[20] Wyatt MA, Lee J, Ahilan Y, Magarvey NA. Bioinformatic evaluation of the secondary metabolism of antistaphylococcal environmental bacterial isolates. Canadian Journal of Microbiology. 2013;**59**:465-471

[21] Khaldi N, Seifuddin FT. SMURF: Genomic mapping of fungal secondary metabolite clusters. Fungal Genetic Biology. 2010;**47**:736-741

[22] Tajkarimi M, Ibrahim S, Cliver D. Antimicrobial herb and spice compounds in food. Food Control. 2010;**21**:1199

[23] Smith-Palmer AJ, Fyfe SL. Antimicrobial properties of plant essential oils and essences against five important food-borne pathogens. Letters in Applied Microbiology. 1998;**26**:118-122

[24] Janjua S, Shahid M, Fakhir-i-Abbas. Phytochemical analysis and in vitro antibacterial activity of root peel extract of *Raphanus sativus* L. var niger. Advancement in Medicinal Plant Research. 2013;**1**:1-7

[25] McCarrell EM, Gould SWJ, Fielder MD, Kelly AF, Sankary WE, Naughton DP. Antimicrobial activities of pomegranate rind extracts: enhancement by addition of metal salts and vitamin C. BMC Complementary and Alternative Medicine. 2008;**8**:64

[26] Mucete D, Borozan A, Radu F, Jainu I, Alexa E. Research about the antimicrobial action of some active principles in Armoracia Rusticana. Agroalimentary Processes and Technology. 2005;**11**(1):237-242

[27] Negi PS, Jayaprakasha GK. Antioxidant and antibacterial activities of *Punica granatum* peel extracts. Journal of Food Science. 2006;**68**(4):1473-1477

[28] Prakash A, Mathur K, Vishwakarma A, Vuppu S, Mishra B. Comparative assay of antioxidant and antibacterial properties of Indian culinary seasonal fruit peel extracts obtained from Vellore, Tamilnadu. International Journal of Pharmaceutical Sciences Review and Research. 2013;**19**:131-135

[29] Roundsa L, Havensa CM, Feinsteinb Y, Friedmanc M, Ravishankar S. Concentration-dependent inhibition of *Escherichia coli* O157:H7 and heterocyclic amines in heated ground beef patties by apple and olive extracts, onion powder and clove bud oil. Meat Science. 2013;**94**:461-467

[30] Chauhan AS, Negi PS, Ramteke RS. Antioxidant and antibacterial activities of aqueous extract of Seabuckthorn (*Hippophae rhamnoides*) seeds. Fitoterapia. 2007;**78**(7-8):590-592

[31] Negi PS, Chauhan AS, Sadia GA, Rohinishree YS, Ramteke RS. Antioxidant and antibacterial activities of various seabuckthorn (*Hippophae rhamnoides* L.) seed extracts. Food Chemistry. 2005;**92**:119-124

[32] Sudarshan S, Fairoze N, Wilfred-Ruban S, Badhe R, Raghunath BV. Effect of aqueous extract and essential oils of ginger and garlic as decontaminant in chicken meat. Research Journal of Poultry Sciences. 2010;**3**:58-61

[33] Indu MN, Hatha AAM, Abirosh C, Harsha U, Vivekanandan G. Antimicrobial activity of some of the south-Indian spices against serotypes of Escherichia coli, Salmonella, *Listeria monocytogenes* and *Aeromonas hydrophila*. Brazilian Journal of Microbiology. 2006;**37**:153-158

[34] Sutherland J, Miles M, Hedderley D, Li J, Devoy S, Sutton K, Lauren D. In vitro effects of food extracts on selected probiotic and pathogenic bacteria. International Journal of Food Sciences and Nutrition. 2009;**60**:717-727

[35] Cicerale S, Lucas LJ, Keast RSJ. Antimicrobial, antioxidant and anti-inflammatory phenolic activities in extra virgin olive oil. Current Opinion in Biotechnology. 2012;**3**(2):129-135

[36] Gutierrez J, Barry-Ryan C, Bourke P. Antimicrobial activity of plant essential oils using food model media: Efficacy, synergistic potential and interaction with food components. Food Microbiology. 2006;**26**:142-150

[37] Harika VC, Padmavathi P, Rao KRSS, Phani RSCH. In-vitro anti microbial activity of leaf powder. International Journal of Research in Pharmaceutical and Biomedical Sciences. 2010;**1**:128-131

[38] Rauha JP, Remes S, Heinonen M, Hopia A, Kähkönen M, Kujala T, Pihlaja K, Vuorela H, Vuorela P. Antimicrobial effects of Finnish plant extracts containing flavonoids and other phenolic compounds. International Journal of Food Microbiology. 2009;**56**:312

[39] Del Nobile MA, Conte A, Cannarsi M, Sinigaglia M. Strategies for prolonging the shelf life of minced beef patties. Journal of Food Safety. 2009;**29**:14-25

[40] Del Nobile MA, Corbo MR, Speranza B, Sinigaglia M, Conte A, Caroprese M. Combined effect of MAP and active compounds on fresh blue fish burger. International Journal of Food Science & Microbiology. 2009;**135**:281-287

[41] Del Nobile MA, Di Benedeto N, Suriano N, Conte A, Corbo MR, Sinigaglia M. Combined effects of chitosan and MAP to improve the microbial quality of *Amaranth* homemade fresh pasta. Food Microbiology. 2009;**26**:587-591

[42] Sofos JN. Challenges to meat safety in the 21st century. Meat Science. 2008;**78**:3-13

[43] Buzby JC, Roberts T. Economic costs and trade impacts of microbial foodborne illness. World Health Statistics Quarterly. 1997;**50**:57-66

[44] Dave D, Ghaly AE. Meat spoilage mechanisms and preservation techniques: A critical review. The American Journal of Agricultural and Biological Science. 2011;**6**:486-510

[45] Jamilah MB, Abbas KA, Rahman RA. A review on some organic acids additives as shelf life extenders of fresh beef cuts. American Journal of Agricultural and Biological Sciences. 2008;**3**:566-574

[46] Cerveny J, Meyer JD, Hall PA. Microbiological spoilage of meat and poultry products. In Compendium of the microbiological spoilage of foods and beverages. Springer New York. 2009:69-86

[47] Janes MJ, Kooshesh S, Johnson MG. Control of *Listeria monocytogenes* on the surface of refrigerated, ready-to-eat chicken coated with edible zein film coatings containing nisin and/or calcium propionate. Journal of Food Science. 2002;**67**:2754-2757

[48] Viuda-Martos M, Ruiz-Navajas Y, Fernández-López J, Pérez-Álvarez JA. Effect of packaging conditions on shelf-life of *Mortadella* made with citrus fibre washing water and thyme or rosemary essential oil. Food and Nutrition Science. 2011;**2**:1-10

[49] Colak H, Hampikyan H, Bingol EB, Aksu H. The effect of risine and bovine lactoferrin on the microbiological quality of turkish-style meatball (Tekirdag Köfte). Journal of Food Safety. 2008;**28**:35-37

[50] Gram L, Dalgaard P. Fish spoilage bacteria—Problems and solutions. Current Opinion in Biotechnology. 2002;**13**:262-266

[51] Huang NY, Ho CP, McMillin KW. Retail shelf-life of pork dipped in organic acid before modified atmosphere or vacuum packaging. Journal of Food Science. 2005;**70**:382-387

[52] Ahmed AM, Ismail TH. Improvement of the quality and shelf-life of minced-beef mixed with soyprotein by Sage (*Salvia officinalis*). African Journal of Food Science. 2010;**4**:330-334

[53] Ayala-Zavala JF, González-Aguilar GA. Optimizing the use of garlic oil as antimicrobial agent on fresh-cut tomato through a controlled release system. Journal of Food Science. 2010;**75**:398-405

[54] Ayala-Zavala JF, Oms-Oliu G, Odriozola-Serrano I, González-Aguilar GA, Álvarez-Parrilla E, Martín-Belloso O. Bio-preservation of fresh-cut tomatoes using natural antimicrobials. European Food Research and Technology. 2008;**226**:47-55

[55] Gutierrez J, Barry-Ryan C, Bourke P. The antimicrobial efficacy of plant essential oil combinations and interactions with food ingredients. International Journal of Food Microbiology. 2008;**124**:91-97

[56] Braga CV, Freire Fuente MF, Freitas ER, de Carvalho LE, de Sousa FM, Bastos SC. Effect of inclusion of coconut meal in diets for laying hens. Revista Brasileira de Zootecnia. 2005;**34**:76-80

[57] Viuda-Martos M, Ruiz-Navajas Y, Fernandez-Lopez J, Perez-Alvarez JA. Effect of added citrus fibre and spice essential oils on quality characteristics and shelf-life of mortadella. Meat Science. 2010;**85**:568-576

[58] Lambert RJW, Skandamis PN, Coote PJ, Nychas GJ. A study of the minimum inhibitory concentration and mode of action of oregano essential oil, thymol and carvacrol. Journal of Applied Microbiology. 2001;**91**:453-462

[59] Mangena T, Muyima NYO. Comparative evaluation of the antimicrobial activities of essential oils of *Artemisia afra, Pteronia incana* and *Rosemarinus officinalis* on selected bacteria and yeast strains. Letters in Applied Microbiology. 1999;**28**:291-296

[60] Marino M, Bersani C, Comi G. Impedance measurement to study antimicrobial activity of essential oils from *Lamiaceae* and *Compositae*. International Journal of Food Microbiology. 2001;**67**:187-195

[61] Jiang Y, Li Y. Effects of chitosan coating on postharvest life and quality of longan fruit. Food Chemistry. 2001;**73**:139-143

[62] Karabagias I, Badeka A, Kontominas MG. Shelf life extension of lamb meat using thyme or oregano essential oils and modifi ed atmosphere packaging. Meat Science. 2011;**88**:109-116

[63] Fu Y, Zu Y, Chen L, Shi X, Wang Z, Sun S, Efferth T. Antimicrobial activity of clove and rosemary essential oils alone and in combination. Phytotherapy Research. 2007;**21**:989-994

[64] Fernàndez-Lòpez J, Zhi N, Aleson-Carbonell L, Pèrez-Alvarez JA, Kuri V. Antioxidant and antibacterial activities of natural extracts: application in beef meatballs. Meat Science. 2005;**69**:371-380

[65] Fernández-Pan I, Royo M, Maté JI. Antimicrobial activity of whey protein isolates edible films with essential oils against food spoilers and food-borne pathogens. Journal of Food Science. 2012;**77**:383-390

[66] Aymerich T, Picouet PA, Monfort, JM. Decontamination technologies for meat products. Meat Science. 2008;**78**:114-129

[67] Del Campo J, Amiot MJ, Nguyen-The C. Antimicrobial effect of rosemary extracts. Journal of Food Protection. 2009;**10**:59-68

[68] Govaris A, Solomakos N, Pexara A, Chatzopoulou PS. The antimicrobial effect of oregano essential oil, nisin and their combination against Salmonella Enteritidis in minced sheep meat during refrigerated storage. International Journal of Food Microbiology. 2010;**137**:175-180

[69] Hyldgaard M, Mygind T, Meyer RL. Essential oils in food preservation: Mode of action, synergies, and interactions with food matrix components. Frontiers in Microbiology. 2012;**3**:3-12

[70] Karaman S, Digrak M, Ravid U, Ilcim A. Antibacterial and antifungal activity of the essential oils of Thymus revolutus Celak from Turkey. Journal of Ethnopharmacology. 2001;**76**:183-186

[71] Min BJ, Han IY, Dawson PL. Antimicrobial gelatin films reduce *Listeria monocytogenes* on turkey bologna. Poultry Science. 2010;**89**:1307-1314

[72] Muthukumarasamy P, Han JH, Holley RA. Bactericidal effects of *Lactobacillus reuteri* and allyl isothiocyanate on *E. coli* O157, H7 in refrigerated ground beef. Journal of Food Protection. 2003;**66**:2038-2044

[73] Nadarajah D, Han JH, Holley RA. Use of allyl isothiocyanate to reduce E. coli O157:H7 in packaged ground beef patties. In: Institute of Food Technology Annual Meeting; Anaheim, CA; 2002; Abstract # 100B-15

[74] Gandhi M, Chikindas ML. Listeria: A foodborne pathogen that knows how to survive. International Journal of Food Microbiology. 2007;**113**:1-15

[75] Lv F, Liang H, Yuan Q, Li C. In vitro antimicrobial effects and mechanism of action of selected plant essential oil combinations against four food-related microorganisms. Food Research International. 2011;**44**:30-57

[76] Moreno S, Scheyer T, Romano CS, Vojnov AA. Antioxidant and antimicrobial activities of rosemary extracts linked to their polyphenol composition. Free Radical Research. 2006;**40**:223-231

[77] Hayouni EA, Chraief I, Abedrabba M, Bouix M, Leveau JY, Mohammed H, Hamdi M. Tunisian *Salvia officinalis* L. and *Schinus molle* L. essential oils: Their chemical compositions and their preservative effects against Salmonella inoculated in minced beef meat. International Journal of Food Microbiology. 2008;**125**:242-251

[78] Krug NS, Cortina D. Myxoprincomide, a novel natural product from Myxococcus xanthus discovered by a comprehensive secondary metabolome mining approach. Angewandte Chemie International Edition English. 2012;**5**:811-816

[79] Liu DC, Tsau RZ, Lin YC, Jan SS, Tan FJ. Effect of various levels of rosemary or Chinese mahogany on the quality of fresh chicken sausage during refrigerated storage. Food Chemistry. 2009;**117**:106-113

[80] Busatta C, Mossi AJ, Alves Rodrigues MR, Cansian RL, de Oliveira J. Evaluation of *Origanum vulgare* essential oil as antimicrobial agent in sausage. Brazilian Journal of Microbiology. 2007;**38**:610-616

[81] Busatta C, Vidal RS, Popiolski AS, Mossi AJ, Dariva C, Rodrigues MRA, Corazza FC, Corazza ML, Oliveira VJ, Cansian RL. Application of *Origanum majorana* L. essential oil as an antimicrobial agent in sausage. Food Microbiology. 2008;**25**:207-211

[82] Lo AH, Liang YC, Lin-Shiau SY, Ho CT, Lin JK. Carnosol, an antioxidant in rosemary, suppresses inducible nitric oxide synthase through down–regulating nuclear factor-κB in mouse macrophages. Carcinogenesis. 2002;**23**:983-999

[83] Careagaa M, Fernandez E, Dorantesa L, Mota L, Jaramillo ME, Hernandez-Sancheza H. Antibacterial activity of capsicum extract against *Salmonella typhimurium* and *Pseudomonas aeruginosa* inoculated in raw beef meat. International Journal of Food Microbiology. 2003;**83**:331-335

[84] Li Y, Guo C, Yang J, Wei J, Xu J, Cheng S. Evaluation of antioxidant properties of pomegranate peel extract in comparison with pomegranate pulp extract. Food Chemistry. 2006;**96**:254-260

[85] Scallan E, Hoekstra RM, Angulo F, Tauxe RV, Widdowson MA, Roy SL, Jones JL, Griffin PM. Foodborne illness acquired in the United States—Major pathogens. Emerging Infectious Diseases. 2011;**17**:7-15

[86] Hitchins AD, Jinneman K. Bacteriological Analytical Manual Chapter 10: Detection and Enumeration of Listeria monocytogenes in Foods. US Food and Drug Administration, 10903; 2011

[87] Zhou GH, Xu XL, Liu Y. Preservation technologies for fresh meat—A review. Meat Science. 2010;**86**:119-128

[88] Gutiérrez L, Batlle R, Andújar S, Sánchez C, Nerín C. Evaluation of antimicrobial active packaging to increase shelf life of gluten free sliced bread. Packaging Technology Science. 2011;**24**:85-94

[89] Petrou S, Tsiraki M, Giatrakou V, Savvaidis IN. Chitosan dipping or oregano oil treatments, singly or combined on modified atmosphere packaged chicken breast meat. International Journal of Food Microbiology. 2012;**156**:264-271

[90] Hecer C, Guldas M. Effects of lactic acid, fumaric acid and chlorine dioxide on shelf-life of broiler wings during storage. African Journal of Microbiology. 2011;**23**:880-883

[91] Oussalah M, Caillet S, Salmiéri S, Saucier L, Lacroix M. Antimicrobial and antioxidant effects of milk protein-based film containing essential oils for the preservation of whole beef muscle. Journal of Agricultural and Food Chemistry. 2004;**52**:5598-5605

[92] Gómez-Estaca J, López de Lacey A, López-Caballero ME, Gómez-Guillén MC, Montero P. Biodegradable gelatin-chitosan films incorporated with essential oils as antimicrobial agents for fish preservation. Food Microbiology. 2010;**27**:889-896

Section 3

Additives from Microbial Origin

Chapter 5

Waste Degradation and Utilization by Lactic Acid Bacteria: Use of Lactic Acid Bacteria in Production of Food Additives, Bioenergy and Biogas

Galina Novik, Olga Meerovskaya and Victoria Savich

Additional information is available at the end of the chapter

http://dx.doi.org/10.5772/intechopen.69284

Abstract

Lactic acid bacteria (LAB) are one of the most well-studied bacterial groups known from ancient times. These valuable microorganisms are used in numerous areas, especially food industry and medicine. LAB produce a wide range of compounds for food upgrading. Moreover, LAB can find special applications like generation of bioenergy not affecting the surrounding environment. The article considers physiological and biochemical processes determining valuable characteristics of the bacteria, potential applications of LAB and their products, especially in food industry and bioenergy sector, and discusses LAB potential contribution into solution of waste disposal problem.

Keywords: lactic acid bacteria, metabolites, waste degradation, food additives, bioenergy

1. Introduction

Lactic acid bacteria (LAB) named so for the appropriate ability to ferment carbohydrates into lactic acid are one of the most studied and used groups of microorganisms. These bacteria have been applied in food processing since ancient times. The first pure culture of LAB was obtained in 1873; however, milk souring and lactic acid producing bacteria were considered as the same microorganisms until the beginning of twentieth century [1]. Today LAB represent a vast and diverse microbial group playing an important role in dairy, baking technology, fish and meat processing. Moreover, LAB are components of normal human microflora and can be used as probiotics to provide health benefits. Thus, LAB find wide applications in food industry and medicine. In addition, it is possible to use the bacteria and their products in other fields, such as generation of bioenergy, wood protection, agriculture, bioremediation of environment and so on.

Lactic acid is the main product of LAB synthesis primarily consumed by food industry. These microorganisms are also sources of low calorie sugars, ethanol, aroma compounds, bacteriocins, exopolysaccharides (EPS) and several vitamins utilized in various areas [2].

LAB can be found in any environment rich in carbohydrates. Waste substrates containing these substances, especially food residues, provide an excellent opportunity for LAB cultivation and fabrication of derived products, cost reduction and refuse disposal. Some carbohydrate compounds can be extracted from wastes, like chitin.

This article presents data on taxonomy, identification, physiology and metabolism of LAB, applications of the bacteria and their products, especially in food industry and contribution in production of bioenergy and biogas.

2. Lactic acid bacteria (LAB): taxonomy and identification, physiological and metabolic processes

LAB represent a diverse microbial group united by the ability to produce lactic acid from various substrates. The first pure culture of LAB, now known as *Lactococcus lactis*, was isolated in 1873 by Lister [1]. Originally the term "lactic acid bacteria" denoted "milk souring organisms," but it came out of use after publication of the monograph by Orla-Jensen (1919) formulating the principles of modern LAB classification [3]. Taxonomic affiliation of the bacteria based on cellular morphology, mode of glucose fermentation, growth temperatures and range of sugar utilization distinguished four core genera: *Lactobacillus*, *Leuconostoc*, *Pediococcus* and *Streptococcus*. The above-mentioned characteristics are still very important for current identification of LAB, despite development of molecular methods. Today LAB are referred to phylum *Firmicutes*, order *Lactobacillales*, genera *Aerococcus*, *Alloiococcus*, *Carnobacterium*, *Enterococcus*, *Lactobacillus*, *Lactococcus*, *Leuconostoc*, *Oenococcus*, *Pediococcus*, *Streptococcus*, *Symbiobacterium*, *Tetragenococcus*, *Vagococcus* and *Weissella*. Sometimes species of *Bifidobacterium* genus (phylum *Actinobacteria*) are assigned to LAB [4]. This genus was distinguished as a separate taxon in 1973 [5]. Until that moment, *Bifidobacterium* was incorporated in other genera, including *Lactobacillus*.

Because of LAB beneficial properties, their correct identification is vital for further industrial and medical use. Phenotypic methods are cheaper compared to genotypic methods, but similar phenotypes displayed by strains do not always correspond to similar or even closely related genotypes. Phenotypic methods also differ by poor reproducibility and ambiguity of some techniques often caused by complex growth conditions, weak discriminatory power and massive arrangements for large-scale studies. Among these methods, protein profiling seems quite reliable for LAB identification. However, even this procedure demands high workload and lacks discriminatory power on the subspecies level, for example, in the *Lactobacillus acidophilus* group [6].

http://dx.doi.org/10.5772/intechopen.69284

In turn, genotypic methods are not dependent on growth conditions of microorganisms and exhibit various levels of differentiation, from species to individual strains (typing), but they are labor-consuming. Sequencing of the 16S rRNA gene is the most popular molecular tool of identification. Some features make this gene an attractive research object: it is present in all bacteria; 16S rRNA function has remained stable over a long period, so random sequence changes reflect measure of time; the gene is large enough (approximately 1500 bp) to contain statistically significant sequence information [7]. Besides sequencing of 16S rRNA gene, it is possible to carry out hybridization of oligonucleotide probes to reveal taxonomic groups with different specificity from domain to species level. In case of intraspecific identification, other methods are practiced. They include DNA fingerprinting techniques: restriction fragment length polymorphism analysis involving the digestion of genomic DNA with restriction enzymes to large fragments fractionated using pulsed-field gel electrophoresis; randomly amplified polymorphic DNA analysis applying arbitrary primers for amplification of corresponding DNA fragments; amplified restriction length polymorphism method combining two previous techniques and so on. All these methods are successfully used in identification and differentiation of LAB [8].

LAB are Gram-positive rods and cocci characterized by the absence of catalase (although some strains can produce pseudocatalase), tolerance to low pH values and lack of spore formation. These bacteria do not synthesize components of respiratory chains such as cytochromes and porphyrins and cannot generate ATP via proton-gradient mechanism. Therefore, LAB produce ATP predominantly by fermentation of sugars. Because of lack of cytochromes and porphyrins, LAB do not use oxygen, but they can grow in its presence. Protection from oxygen by-products (e.g. H_2O_2) is provided by peroxidases [9].

The distinctive feature of LAB is production of lactic acid. They are chemotrophic microorganisms deriving necessary energy from oxidation of chemical compounds, especially sugars. There are two fermentation pathways: homofermentative and heterofermentative. Homofermentative bacteria produce lactic acid as the major metabolite through glycolysis or Embden-Meyerhof-Parnas pathway generating two moles of lactate per mole of glucose. Pentoses and gluconate are not fermented by microorganisms via obligate homofermentative pathway due to lack of enzyme phosphoketolase. This type of fermentation is inherent, for example, to some species of the genus *Lactobacillus* (*L. acidophilus, L. delbrueckii, L. helveticus, L. salivarius*).

In turn, heterofermentative microorganisms using pentose phosphoketolase pathway (hexose monophosphate shunt/6-phosphogluconate pathway) produce equimolar amounts of lactate, CO_2 and ethanol (**Figure 1**). Genera *Leuconostoc, Oenococcus, Weissella* and some lactobacilli (*L. brevis, L. buchneri, L. fermentum, L. reuteri*) are characterized by this type of fermentation. Hexoses other than glucose enter the major pathways after different isomerization and phosphorylation steps [10].

Genus *Bifidobacterium* differs from LAB by alternative way of sugar conversion known as *bifid shunt*. Hexoses are degraded through several stages to acetyl-phosphate 2-glyceraldehyde-3-phosphate. The latter is metabolized by Embden-Meyerhof-Parnas pathway to lactic and

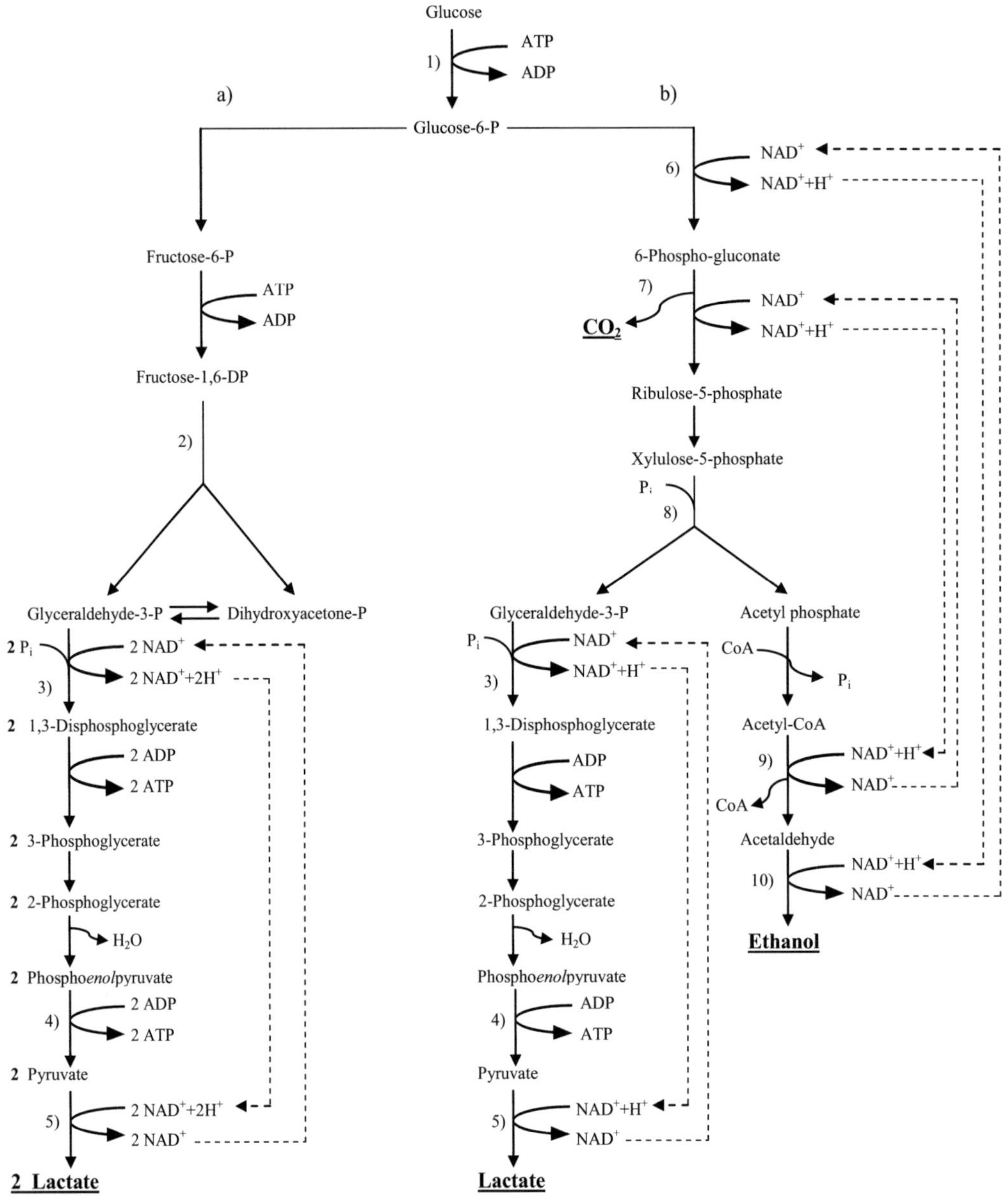

Figure 1. (a) Homefermentative and (b) heterofermentative pathways of lactic acid production. Key enzymes: (1) glucokinase; (2) fructose-1,6-diphosphate aldolase; (3) glyceradehyde-3-phosphate dehydrogenase; (4) pyruvate kinase; (5) lactate dehydrogenase; (6) glucose-6-phosphate dehydrogenase; (7) 6-phospho-gluconate dehydrogenase; (8) phosphoketolase; (9) acetaldehyde dehydrogenase and (10) alcohol dehydrogenase.

acetic acid in the ratio 2:3. This pathway yields 2.5 moles ATP per mole of glucose, whereas homofermentative lactic acid fermentation generates 2 moles of ATP per mole of glucose [11].

Many LAB are able to ferment pentoses. They can digest them heterofermentatively by entering the phosphogluconate pathway as either ribulose-5-phosphate or xylulose-5-phosphate. Pentoses are converted into lactate and acetate, with no CO_2 evolved [12]. Disaccharides are previously split enzymatically into monosaccharides that enter the appropriate pathways [11].

The proteolytic system of LAB converts proteins to peptides and then to amino acids essential for bacterial growth. The branched-chain amino acids (valine, leucine and isoleucine), the aromatic amino acids (tyrosine, tryptophan and phenylalanine) and the sulfur-containing amino acids (methionine and cysteine) are the main amino acid sources for flavor compounds, such as aldehydes, alcohols and esters, generated using two distinct routes: transamination and elimination [13]. The proteolytic system of LAB includes three major components: cell-wall bound proteinase initiating degradation of extracellular milk protein casein into oligopeptides; transporters taking up the peptides into the cell and various intracellular peptidases degrading the peptides into shorter fragments and amino acids. Components of the proteolytic system are diverse in various groups of LAB as well as in distinct strains. Some enzymes are only found in a few LAB strains, such as cell-wall bound proteinase PrtP. Other ferments, like aminopeptidases PepC, PepN and PepM, and proline peptidases, PepX and PepQ, are represented in all genomes, usually with one gene per genome. It appears logical that bacteria with extensive set of proteolytic enzymes show certain advantages when applied in manufacturing of various compounds [14].

Lipid metabolism proceeds as the breakdown of lipids by lipases into fatty acids and glycerol. LAB are able to produce lipases, but they are less efficient if compared with other microorganisms, such as *Pseudomonas, Aeromonas, Acinetobacter* or *Candida,* and mostly intracellular. Besides, not all LAB synthesize these enzymes. Only one quarter of lipase-producing strains were detected among 103 tested LAB from the genera *Lactobacillus, Lactococcus, Leuconostoc, Pediococcus* and *Streptococcus*. The majority belonged to *Lactococcus* species [15]. LAB can perform unique fatty acid transformation reactions: isomerization, hydration, dehydration and saturation. Some products of lipid metabolism, e.g. conjugated linoleic acid, can be used for medicinal and nutraceutical purposes [16]. Esterases of LAB are able to catalyze both hydrolysis of fat glycerides with release of free fatty acids and ester synthesis from glycerides and alcohols via transferase reaction. Esterases display the highest activity on monoglycerides, with inferior activity on diglycerides. However, their activities on the specific glycerides and ρ-nitrophenyl or β-naphthyl esters of fatty acids decrease as the carbon-chain length of the esterified fatty acid increases [17].

3. Applications of LAB

LAB are applied in food production and preservation from the ancient times. Nowadays, LAB find wide use in various areas such as synthesis of chemicals and pharmaceuticals or manufacturing of probiotics for agriculture and medicine. Nevertheless, food industry remains to be the domain of broad LAB application. LAB strains were granted "Qualified Presumption of Safety" and "Generally Regarded as Safe" status by the European Food Safety Authority (EFSA) and Food and Agriculture Organization of the United Nations (FAO), respectively. They are used in manufacturing of dairy, meat, baking and vegetable products all over the world [18–21]. These bacteria also allay product allergenicity and ensure longer preservation of fermented foods [22]. LAB can be involved in the delivery of functional biomolecules and ingredients into high quality gluten-free cereal products [23]. In the seafood industry, LAB are usually applied for product conservation, with the exception of traditional fish sauces in Southeast Asia. In recent years, novel fish products with various flavor and biochemical characteristics have been developed [24, 25].

Another direction of LAB application is beverage production. LAB are important components of the wine-making process: they are responsible for malolactic fermentation following alcoholic fermentation by yeast. Nearly all red wines and many white wines are obtained by these two fermentation steps. When all reducing sugars are converted to ethanol, yeast concentration declines and LAB start to grow consuming residual sugars and transforming numerous wine components. New aromas may improve wine bouquet, whereas those revealed during alcoholic fermentation by yeast are likely to vanish or change after malolactic fermentation. Some strains of LAB could even spoil wine during the process [26].

LAB are part of normal microflora of gastrointestinal and genitourinary tracts, hence they are used as components of probiotics. Beneficial effects of probiotics are provided by several mechanisms. Antagonistic action toward pathogenic bacteria may be manifested by decreasing the luminal pH through production of volatile short-chain fatty acids (SCFA), such as acetic, lactic or propionic acid; rendering specific nutrients not digestible by pathogens; decreasing the redox potential of the luminal environment; producing hydrogen peroxide under anaerobic conditions and specific inhibitory compounds such as bacteriocins affecting other bacteria [27, 28]. Besides the above-mentioned synthesis of various compounds, probiotics can be engaged in barrier function, modulation of the mucosal immune system, enhancement of food digestion and absorption and alteration of the intestinal microflora [29].

LAB can be used to control a wide range of diseases: diarrhea of various etiology [30], allergy [31–33], inflammatory bowel diseases [34] and hepatic diseases [35]. LAB are applied in the treatment of tumors such as colorectal cancer by several mechanisms: bacteria are able to cause apoptosis of tumor cells; they possess antioxidative activity; LAB stimulate immune response for cancer prevention and therapy; they are able to modify expression levels of selected genes and LAB suppress proliferation of cancer cells via synergistic action of adherence to tumor cells and production of SCFA [36]. Some LAB display cholesterol-lowering and antihypertensive effects and alleviate the symptoms of lactose intolerance in lactase-deficient individuals [37–40]. LAB were shown to promote immunomodulatory impact on human organism [41, 42].

LAB facilitate target delivery of valuable substances. Selenium is an essential trace element that protects organism from oxidative stress, helps maintain defense barrier against infections, modulates growth and development, provides for normal aging process, minimizes pregnancy complications and improves fertility and antiviral activity. Selenium-enriched probiotics have been shown to confer several health benefits on the host due to their antioxidative, antipathogenic, antimutagenic, anticancerogenic and anti-inflammatory activities [43].

LAB can be applied in prevention and treatment of animal diseases. Viruses, such as the infectious pancreatic necrosis virus and infectious hematopoietic necrosis virus, cause acute diseases of rainbow trout (*Oncorhynchus mykiss*) and several salmon species. The purified dextrans of

Lactobacillus sakei MN1 and *Leuconostoc mesenteroides* RTF10 have shown functional activity against these viruses [44]. In some cases, *Enterococcus* strains demonstrated prophylactic and therapeutic effect and stimulated immune response, growth and digestion in farm stock and pets [45]. Several studies testing the influence of various LAB species on pigs, poultry and ruminants established the elevated titer of beneficial bacteria and the reduction of potential pathogen load [46].

LAB and their products exhibit antifungal properties applicable in agriculture, food and wood industry. Fungi cause numerous diseases of crops and decrease yields. In addition, they impart an unpleasant smell, taste or appearance to feed and foodstuffs and produce a wide array of mycotoxins, making nutriment unsuitable for consumption. They cause adverse effects up to lethal cases after penetration into human or animal body [47]. LAB are able to inhibit fungal growth and to dispose of mycotoxins. The activity of LAB can be explained by synthesis of various compounds, competition for nutrients in the medium and/or acidification of the growth medium. Detoxification capacity can be related to adsorption of mycotoxins by the bacterial cell [48–50]. Even heat-killed cells of LAB may reduce toxin concentrations to safe levels in milk. Heat inactivation significantly enhanced aflatoxin M_1 removal by LAB [51]. Members of genera *Lactococcus, Pediococcus, Leuconostoc* and *Lactobacillus* are the most promising bacteria to inhibit fungal growth [50, 52, 53]. Both lactococci and yeast could delay or prevent the fungal deterioration of the baked food [53]. A multitude of studies showed LAB ability to block fungal spoilage of fresh fruits and vegetables, baked and dairy products and silage [54]. Besides, LAB were shown to inhibit the growth of wood-rotting fungi and subsequent wood decay [55].

1,3-Propanediol is a monomer in polymerization process producing polytrimethylene terephthalate, and it can also be used in the production of polyurethanes, polyesters and polyethers. A large number of microorganisms, including LAB, are capable of converting glycerol into 1,3-propanediol. The 1,3-propanediol concentration achieved in batch cultivation of *Lactobacillus diolivorans* equalled 41.7 g/L. This value could be increased to 84.5 g/L by co-feeding glucose and glycerol (in 0.1 molar ratio) and by adding vitamin B_{12}, the co-factor of glycerol dehydratases [56]. Recent studies have revealed possibility of applying LAB in biosurfactant production. Biosurfactants are a structurally diverse group of surface-active substances used in agriculture, food production, chemistry, cosmetics and pharmaceutics [57].

LAB potential application area is bioremediation, e.g. treatment of wastewaters containing azo dyes. The latter make up the largest group of synthetic chemicals that are widely used in manufacturing of textile, leather, cosmetics, food and paper. During the industrial process, approximately 10–15% of the spent dye is discharged into wastewater. Azo pigments and their catabolic intermediates, like aromatic amines, distinguished by mutagenic and carcinogenic properties, obstruct light and oxygen transfer into water bodies, consequently affecting aquatic life. The research data indicate that the chemical can be catabolized and utilized by LAB strains and its degradation products are less toxic to growing *Sorghum bicolor* culture than the original azo pigment [58].

LAB and their products can be used for crude oil recovery. One third to a half of the world oil reserves are deposited in carbonate rock. They tend to have very low permeability that can be improved by acid injection. Microbial acid producers, like LAB, may provide a solution for the problem. They are injected with nutrient substrate into the well where bacteria produce lactic acid reacting with $CaCO_3$. The water solubility of formed calcium lactate is approximately 80g/L as compared to 15mg/L for $CaCO_3$. Lactic acid may also be used for the removal of carbonate or iron scale from oilfield equipment [59].

4. Waste degradation and utilization by lactic acid bacteria

One-third of food intended for human consumption is lost or wasted globally at all steps from initial agricultural production to final household consumption. It amounts to about 1.3 billion tons per year [60]. Food wastes are mainly composed of carbohydrate polymers, such as starch, cellulose and hemicelluloses, plus lignin, proteins, lipids, organic acids and inorganic remainder. Total sugar and protein contents are in the range of 35.5–69 and 3.9–21.9%, respectively [61]. LAB may grow in any environment rich in carbohydrates, so that they can be found in various food products (milk, meat and vegetables), plants, as part of the normal human and animal microbiota. Food wastes are potential sources of nutrients for growth of LAB and production of valuable compounds.

Large volumes of waste generated by fishing, aquaculture or food processing are dumped into the sea without pretreatment. It causes grave environmental problems. This challenge can be met by introducing rich organic nutrients in the formulated optimum media for microbial cultivation. Enzymatic hydrolysate of octopus processing wastewater served as a good source for LAB growth (*L. lactis* and *Pediococcus acidilactici*) and synthesis of bacteriocins (nisin and pediocin, respectively). The maximal production of biomass and nisin by *L. lactis* was observed in the media with low concentration of enzyme papain and short time of hydrolysis (4 h). In case of pediocin, the highest production was attained in the media hydrolyzed with papain, trypsin and pepsin within 10 h period. Consequently, marine peptones are promising alternative nutrients in the media and their fermentation is a possible solution of wastewater problem [62]. Fish viscera waste can be used in preparation of silage intended as animal feed. Application of LAB makes bio-silage process simpler, faster, more environmentally friendly and cost-efficient than chemical technology. LAB strains produce metabolites and adjust pH values for bio-silage fermentation and preservation [63].

Brown juice, waste of the green crop drying industry, contains nutrients such as carbohydrates, organic acids, vitamins and minerals suitable for production of L-lysine. Pretreatment is required to convert brown juice into a stable, storable product that can be used for microbial fermentation. Traditional heat sterilization at 121°C for 20 min in batch procedure or at 140°C for a few seconds in continuous process inactivates valuable enzymes and consumes a lot of energy. When LAB deplete the constituent carbohydrates, the juice can be heat sterilized and used as a nutrient and water source for L-lysine production by *Corynebacterium* after addition of a carbon source and neutralization of the lactic acid by, e.g., ammonia. Alternatively, the lactic acid present in the medium can be utilized by *Corynebacterium* and converted to L-lysine [64].

LAB can be used for waste preservation. Fermentation of hatchery wastes, including infertile eggs, dead embryos, cull chicks and shells from hatched chicks, by bacteria *Pediococcus acidilactici* and *Lactobacillus plantarum* and products of *Streptococcus faecium* M74 exerted significant effects upon nutritional composition of the treated substrate. Additionally, LAB action reduces or eliminates pathogenic bacteria such as *Salmonella* species and *Escherichia coli*. These are important steps in recycling hatchery by-products into feed ingredients instead of landfilling waste [65]. Rations with fermented hatchery wastes showed no negative effect on broiler chicken. Their body weight gain and feed conversion at all stages were comparable to the control. In some cases, the parameters such as ready to cook carcass and wing yield significantly exceeded control values [66].

Lactic acid is the main product of LAB. The use of waste substrates for production of lactic acid by LAB is described in Section 6.

5. Food additives. Waste for the production of chitin and chitosan

Food additive is any substance added to food to improve its quality. These compounds are used in production, processing, treatment, packaging, transportation or storage of food. Food additives are applied to secure safety and freshness of products that could be spoiled by environment and microorganisms, to upgrade food nutritional value or modify taste, texture and appearance of consumable products. LAB are known to promote food quality and flavor from ancient times, but they also produce specific beneficial compounds that can be used for food supplementation or for extraction of valuable substances such as chitin.

Microbial contamination poses serious safety and quality problems in food industry. Bacteriocins are antimicrobial peptides produced by bacteria, which possess the ability to kill or inhibit other bacteria. The bacteriocins were first characterized in Gram-negative bacteria, but later they were observed in other bacterial groups, including LAB. These compounds are often confused with other antimicrobials or antibiotics. Unlike most antibiotics, which are secondary metabolites, bacteriocins are usually ribosomally synthesized and sensitive to proteases, whereas generally harmless to the human body and surrounding environment. Besides, bacteriocins have narrower spectrum of activity opposite to antibiotics. Bacteriocins are generally divided into several classes. Class I, or the lantibiotics, are small (<5 kDa) thermally stable peptides that contain lanthionine, methyllanthionine and dehydrated amino acids. Subclass Ia are linear structure peptides with membrane-disrupting mode of action, and subclass Ib are globular structure peptides with cellular enzymatic action. Class II containing small (<10 kDa) heat-stable, unmodified non-lanthionine membrane-active peptides is subdivided into five subclasses. Subclass IIa are pediocin-like *Listeria*-active peptides with a consensus amino acid sequence Tyr-Gly-Asn-Gly-Val-Xaa-Cys in the N-terminal position. Subclass IIb consists of two different unmodified peptides forming a fully active poration complex. Subclass IIc are circular peptides. Sublass IId are linear, non-pediocin-like, single-peptide bacteriocins and subclass IIe bacteriocins are non-ribosomal siderophore-type post-translation modification peptides with the serine-rich carboxy-terminal region. Class III bacteriocins are large molecular weight (>30 kDa), thermally unstable proteins that can be further subdivided into two distinct groups with respect to cell lysis. Class IV forms large complexes with other macromolecules [67–69]. Due to sensitivity to proteases, bacteriocins are probably digested in the gastrointestinal tract into small peptides and amino acids. Since bacteriocin-producing bacteria are present in many types of food since ancient times, bacteriocins are considered as basically safe food additives [67]. The main perspective for these compounds is food preservation. There are many studies regarding the role of bacteriocins in conservation of dairy, meat, seafood and vegetable products [70–73]. However, only few bacteriocins are used as commercial biopreservatives. The most well-studied and used bacteriocin is nisin, first isolated from *L. lactis* ssp. *lactis* in 1928 [74]. Nisin approved as food additive in more than 50 countries, including USA and Europe, is marketed as Nisaplin®. It was included into the European food additive list under the number E234 with no recorded adverse effects. Nisin inhibits closely related species as well as food-borne pathogens such as *Listeria monocytogenes* and many other Gram-positive spoilage microorganisms [70]. Another commercially available bacteriocin is pediocin PA-1, marketed as Alta® 2341, which inhibits growth of *L. monocytogenes* [72].

Exopolysaccharides (EPS) of LAB are branched, repeating units of sugars or sugar derivatives produced extracellularly. They are involved in the protection of bacteria from adverse factors. EPS of LAB are versatile in molecular weight, linkages, solubility and degree of branching. The molecular mass of EPS ranges from 10 to 1000 kDa. Most LAB produce polysaccharides extracellularly from sucrose by glycansucrases or intracellularly by glycosyltransferases from sugar nucleotide precursors [75]. These compounds are widely applied in food industry as adjuvants, emulsifiers, carriers, stabilizers, sweeteners, bulking agents, extenders and so on. [76, 77]. EPS of LAB also find use in medicine. They prevent blood coagulation and facilitate blood flow, tissue transfer, tumor treatment, serve as lubricants, carriers, osmotic and hypocholesterolemic agents, etc. [77].

Low calorie sugars of LAB origin are recognized as vital ingredients in diabetic foodstuffs. Mannitol, sorbitol, xylitol, erythritol and D-tagatose are sweeteners produced by LAB. Mannitol is used as a sweet-tasting bodying and texturing agent. It retards sugar crystallization and is intended to increase the shelf life of foods. Crystalline mannitol exhibiting very low hygroscopicity is indispensable in products that keep stability at high humidity. The polyol is usually manufactured by high pressure hydrogenation of fructose/glucose mixtures; however, bacteria can also be used as sources of the compound. *Lactobacillus intermedius* B-3693 was shown to yield mannitol from fructose. For example, 0.70 g of mannitol per gram of fructose can be produced from 250 g/L fructose. It was established that one-third of fructose could be replaced by glucose, maltose, galactose, mannose, raffinose or starch with glucoamylase, or two-thirds of fructose could be replaced by sucrose for successful mannitol production [78]. D-tagatose can be used as a low-calorie sweetener. The sweetness profile of D-tagatose is similar to that of sucrose, but it is detected a bit sooner than that of sucrose. D-tagatose is catabolized via tagatose-6-phosphate pathway, a branch of galactose metabolism, by some microorganisms such as *Lactobacillus casei* and *L. lactis*. L-arabinose isomerase used in tagatose production was found in *L. plantarum* and *Bifidobacterium longum* [79]. Sorbitol is a low-calorie sugar alcohol widely used in food industry. This polyol has a relative sweetness of around 60% when compared to sucrose and displays 20 times higher solubility in water than mannitol. Sorbitol is applied as sweetener, humectant, texturizer and softener in production of chewing gum, candies, desserts, ice cream and diabetic food. *L. plantarum* produces sorbitol with efficiency 61~65% from fructose-6-phosphate by reverting the sorbitol catabolic pathway in a mutant strain deficient for both L- and D-lactate dehydrogenase activities [80]. D-xylitol is a 5-carbon polyol used as a natural sweetener in food and confectionary industry and known for its anticariogenic properties. The recombinant strain *L. lactis* was able to produce D-xylitol during cometabolism of glucose and D-xylose. Xylitol synthesis reached productivity 2.72 g/L/h [81]. *Oenococcus oeni* has been reported to produce erythritol. This polyol is another compound that can be used as sugar substitute [82].

Antioxidant is the compound inhibiting oxidation of other molecules by free radicals. Although synthetic antioxidants are more effective, natural antioxidants are characterized by simpler structure, higher stability and safe immune response. Substances with potential antioxidant activity have been derived from many animal and plant sources. LAB products also show this kind of activity [83]. Some studies demonstrated LAB contribution in production of peptides showing antioxidant activity, with potential food and pharmaceutical applications [84, 85]. However, further investigations are required to evaluate prospects of peptides.

Vitamins are substances essential for metabolic processes. They regulate biochemical reactions in the cell. Some of them function as precursors of coenzymes. Humans are incapable of synthesizing most vitamins, so that they have to be provided from food or synthesized by gut microflora. Regretfully, vitamins are easily degraded during food processing or cooking. Certain strains of LAB possess the property to synthesize vitamins and hence can be engaged in elaboration of enriched fermented foods. Studies indicated LAB production of B-group vitamins and vitamin K [86, 87].

Conjugated linoleic acid (CLA) isomers are other compounds with important physiological properties. CLA represent the family of octadecadienoic acid (18:2) isomers, which have a pair of conjugated double bonds along the alkyl chain. There are 28 known CLA isomers. They are characterized by anticancer, antidiabetic, antiatherosclerotic and anti-osteoporosis activities, complemented by defattening and immune-stimulating functions. The use of LAB and *Bifidobacteria* allows to increase CLA content of fermented dairy products, with no adverse effects described to date. Attempts to raise CLA productivity of LAB have been reported [88].

Apart from nutrient balance, a key food characteristic is flavor. Consumers need not only healthy but also delicious food. LAB showed ability to degrade phenolic acids generating compounds responsible for aroma. Phenolic compounds are directly related to sensory food characteristics such as flavor, astringency and color. In addition, they show antioxidant activity [89]. LAB metabolize phenolic acids by decarboxylation and/or reduction. The products of phenolic acid decarboxylase action are vinylcatechol, vinylphenol, vinylguaiacol, pyrogallol and catechol; reduction of hydroxycinnamic acids yields dihydrocaffeic and dihydroferulic acids [90–92]. Strains with high enzymatic activities can be used to enhance the flavor of cheeses [93]. The volatile flavor components, which predominantly determine the typical odor of cheese, are subsequently derived from the activity of amino acid converting enzymes [94].

Chitin is a polysaccharide composed of N-acetyl-D-glucosamine units. It is the second most abundant biopolymer on Earth after cellulose and it is a structural component of the arthropod exoskeleton and of the cell walls of algae, fungi and yeast. Chitin is the source of chitosan, polysaccharide with numerous applications in the area of food and nutrition, in agriculture and environmental protection, medical, dietetic and cosmetic products. Chitin is widely used to immobilize enzymes and whole cells further engaged in clarification of fruit juices and processing of milk [95]. Chitosan and its derivatives can be applied as thickeners and stabilizers for sauces, fungistatic and antibacterial coating for fruit, preservatives, dietary fibers and cholesterol reducers [96]. Chitin and chitosan are non-toxic compounds displaying excellent biological properties such as biodegradation in the human body, immunological, antibacterial and wound-healing activity [97–100]. They also possess chelating ability and adsorption capacity and promote disposal of unwanted substances or extraction of valuable compounds [101]. Derivatives such as chitosan-sugar complexes show the potential to act as better antimicrobial and antioxidant agents than chitosan itself. Antimicrobial activity of chitosan is displayed by several mechanisms. The available amino group in chitosan structure provides for absorption of the nutrients necessary for bacterial growth. Interaction between the positive charge of chitosan molecule and the negative charge of microbial cell membrane changes membrane permeability. Chitosan film formation over the surface of microbial cell membrane prevents the nutrients from getting into the cell [102].

Chitin is associated with proteins, lipids, pigments and mineral deposits. Therefore, chitinous materials have to be pretreated to remove by-components. Chitin can be extracted by various ways, including LAB introduction. However, demineralization and deproteinization of the chitinous material depend primarily on fermentation conditions. Ninety-one percent of deproteinization with lower level of demineralization can be reached under optimal conditions by *L. helveticus* using date juice as an alternative to glucose that decreased the degree of deproteinization to 76% [103]. The other strain *Pediococcus acidolactici* CFR2182 carried out efficient fermentation of shrimp waste resulting in 97.9% deproteinization and 72.5% demineralization [104]. The epiphytic *L. acidophilus* SW01 culture isolated from shrimp waste quickly removed minerals and proteins from that substrate to residual 0.73 and 7.8% values, respectively, after 48 h fermentation. In the pilot scale fermentation, the mineral and protein contents fell to 0.98 and 8.44%, respectively, after 48 h fermentation [105]. The combination of lactic acid bacteria (*Lactobacillus paracasei*) and protease-producing bacteria (*Serratia marcescens*) can also be effective for extraction of chitin. LAB intensely dissolved mineral $CaCO_3$ by producing organic acid and *S. marcescens* degraded proteins by producing extracellular proteases. The extent of demineralization reached the highest mark of 97.2%, but the percentage of deproteinization in cofermentation was 52.6% on day 7 due to unfavorably low pH for proteolytic activity [106]. Mixed cultures of *L. lactis* and *Teredinobacter turnirae* displayed splendid activity in mineral and protein removal, respectively, and promoted chitin extraction, especially when *T. turnirae* was first inoculated [107]. LAB can recover chitin with accessory compounds such as pigment astaxanthin reported to be an excellent antioxidant and anticarcinogenic substance [108]. Microbial method is more effective for isolation of chitin when compared with chemical method. Adding $Fe(NO_3)_3$ as extra nitrogen source increases yield twice. Organic acids, like lactic acid, can be produced at low cost by bacteria, are less harmful to the environment and can preserve characteristics of the purified chitin, whereas the organic salts from demineralization process can be used as environmentally friendly deicing agents or as preservatives [109].

6. Lactic acid: use of waste substrates for production of lactic acid by LAB

Lactic acid, or 2-hydroxypropanoic acid, is water soluble and highly hygroscopic organic acid with ubiquitous distribution in nature. Lactic acid was discovered in 1780 by C.W. Scheele in sour milk, and in 1881 Fermi obtained this compound by fermentation, resulting in its industrial production. Lactic acid is widely used in food, pharmaceutical, cosmetic and other manufacturing sectors. In the chemical industry, lactic acid is treated as a raw material for production of lactate ester, propylene glycol, 2,3-pentanedione, propanoic acid, acrylic acid, acetaldehyde and dilactide. This compound can even be used for fabrication of polylactic acid (PLA), sustainable bioplastic material mainly applied in packaging. Lactic acid functions as a descaling agent, pH regulator, neutralizer, chiral intermediate, solvent, humectant, cleaning aid, slow acid-release, metal complexing and antimicrobial agents. Technical-grade lactic acid is used in leather tanning industry as an acidulant for deliming hides. Besides moisturizing and pH adjusting effect, the substance is characterized by antimicrobial activity, skin lightening and hydrating action in cosmetic industry. In medicine, lactic acid is applied in tableting, prostheses, surgical sutures, controlled drug delivery systems and electrolyte solutions [110]. However, food industry is the main consumer of lactic acid. Food and food-related applications account for approximately 85% of lactic acid demand, whereas the other industrial sectors cover the remaining 15% [111]. Lactic acid and its salts are used as antimicrobials, flavor enhancers, stabilizers, thickeners, humectants, emulsifiers, firming and leavening agents and so on [110, 112]. Lactic acid is applied in a wide variety of foodstuffs, such as candies, bread and bakery products, soft drinks, soups, sherbets, dairy products, beer, jams and jellies, mayonnaise and processed eggs [113].

The global lactic acid demand estimated to be 714.2 kilo tons in 2013 is expected to reach 1960.1 kilo tons by 2020 [114]. Substrates for lactic acid production should be characterized by cheapness, low contamination level, year-round availability, rapid fermentation rate and high yields of lactic acid from fermentation.

Food waste has high starch content and is rich in nutrients, including lipids and proteins, and therefore it represents a potential renewable resource for lactic acid production. Additionally, protease, temperature and $CaCO_3$ cause significant linear effects on production, whereas α-amylase and yeast extract show minor effects. Under the optimal conditions, *L. plantarum* produced maximum amount of lactic acid from dining hall food waste [115]. Municipal organic solid waste (MOSW) is the discharge consisting of kitchen and garden residues. MOSW possesses high energy and nutritional value for lactic acid production. Lactic acid productivity after 24 h was 0.79 ± 0.05 g/L/h in fermenters with pH 5.0 and 0.71 ± 0.05 g/L/h in fermenters with uncontrolled pH [116].

Sugarcane juice containing 13–16% sucrose is renewable, abundant and cheap source of carbon for lactic acid production. *Lactobacillus* sp. strain FCP2 grown on sugarcane juice for 5 days produced 104 g/L lactic acid with 90% yield. Higher yield (96%) and productivity (2.8 g/L/h) were obtained when the strain was cultured on 3% w/v sugarcane juice for 10 h. Addition of cheap nitrogen sources such as silk worm larvae, beer yeast autolysate and shrimp waste led to increase in lactic acid production over that attained with yeast extract [117]. Molasses is the by-product of refining sugarcane or sugar beet. It contains sucrose (31%), glucose (9.5%), fructose (10%), nitrogen (0.95%) and may be used as cheap and available medium for production of various compounds, including lactic acid. *L. delbrueckii* mutant Uc-3 in batch fermentation process produced 166 g/L lactic acid from 400 g/L molasses [118]. Lactic acid concentration 134.9 g/L was recorded at molasses concentration 333 g/L using *Enterococcus faecalis* culture [119]. Glycerol is the main by-product of biodiesel industry and it can be utilized as a carbon source to yield organic acids, e.g. lactic acid. Strain *Lactobacillus* sp. CYP4 produced 39.41 mM lactic acid with conversion percentage 39.27% [120]. Liquid waste from potato processing industry (chips manufacturing) can be used as substrate for lactic acid production. Waste with MRS medium (lacking peptone, yeast extract and glucose, but containing malt extract, galactose and manganese sulfate) in 4:1 ratio provided for 16.09 g/L concentration of lactic acid by *L. casei* culture [121].

Brewers' spent grain (BSG) represents the major by-product of brewing industry accounting for about 85% of total residues left after the mashing and lautering processes and it is available in large amounts all year around [122]. Chemical composition of BSG varies depending on the barley variety, the harvest time, malting and mashing conditions; however, its hydrolysates are suitable substrates for lactic acid production. Generation of the desired metabolite through fermentation of hydrolysate resulting from BSG pretreatment with aqueous ammonia was 96% higher than that following acid-alkaline treatment and constituted 17.49 g/L. The maximum value was obtained after addition of nitrogen source (yeast extract) to aqueous ammonia-treated BSG (22.16 g/L) [123]. Additional use of invertase from grape juice for sucrose hydrolysis of canned pineapple syrup, a food processing waste, resulted in lactic acid concentrations 20 and 92 g/L generated by *L. lactis* from 20 and 100 g total sugars/L [124].

About 30% of annual global cheese whey production remains underutilized, ending up as waste or animal feed [125]. Besides, most dairy manufactures do not have proper treatment systems for whey disposal. The main components of whey are lactose (approximately 70–72% of the total solids), whey proteins (approximately 8–10%) and minerals (approximately 12–15%) utilized by LAB with lactic acid production. Various studies with free and immobilized cells proved efficiency of LAB application [126]. Scotta is the main by-product of ricotta cheese production containing proteins (0.15–0.22%), salts (1.0–1.13%) and lactose (4.8–5.0%). Scotta may be considered as a source of lactose and other nutrients with potential biotechnological applications such as lactic acid production. The addition of nutritional supplements to medium with scotta led to lactic acid productivity about 2 g/L/h. The use of mixed cultures reduces the need for nutrient supply, with no detrimental effects on the production parameters as compared to pure cultures [127]. Mussel processing waste, liquid by-product of industrial steam treatment of mussels, contains glycogen as the main component that can be utilized by LAB with protein and phosphorus supply and pH control [128].

Deficiency of the nitrogen source usually decreases yield of lactic acid. Moreover, nitrogen source is the most expensive component of microbial growth media. Ram horn hydrolysate (RHH) was shown to be rich in both organic and inorganic compounds and hence considered as an excellent source of nitrogen and minerals in fermentation medium because of its amino acid and mineral contents. The optimal concentration of RHH for production of lactic acid was 6%. Concentrations higher than 6% had an inhibitory effect due to high amounts of heavy metals. 44 g/L concentration of lactic acid was generated on medium with RHH by 26 h of fermentation with nearly 100% sugar consumption in contrast to control medium (36 g/L and the degree of sugar consumption 82%) [129].

Experiments with production of lactic acid were performed on pineapple juice waste [130], waste potato starch [131], cassava powder [132], waste banana [133], kitchen waste [134] and fish waste [135].

Lignocellulosic hydrolysates also can be used for lactic acid production. Lignocellulosic biomass, organic material of biological origin, represents the most abundant global source of unutilized biomass. Lignocellulosics are typically composed of cellulose (insoluble fibers of β-1,4-glucan), hemicellulose (noncellulosic polysaccharides, such as xylans, mannans and glucans) and lignin (a complex polyphenolic structure) with lesser amounts of minerals, oils and other components. The proportion of biomass constituents varies among species. LAB are not able to digest these components, therefore, pretreatment and enzymatic hydrolysis stages are essential [136]. For example, dilute acid pretreatment efficiently hydrolyzes hemicellulose to xylose, arabinose and glucose and thereby enables further enzymatic digestion of cellulose to glucose. The obtained compounds are utilized by LAB. However, substances toxic to fermentative organisms such as furfural, phenolic derivatives and inorganic acids are also produced during the pretreatment process. Strains S3F4 (*L. brevis*) and XS1T3-4 (*Lactobacillus plantrum*) exhibited the ability to utilize various sugars present in dilute-acid hydrolysates of corn stover and corncobs, especially S3F4 converting hydrolysates into lactic acid without detoxification. The strain showed strong resistance to the potential inhibitors, furfural, and ferulic acid. The maximum lactic acid concentration achieved by S3F4 fermentation was 39.1 g/L from corncob hydrolysate [137].

The food processing industry generates significant amounts of solid wastes. For example, over 50% of the orange fruit is transformed into peel waste during the juice making process [138]. Food processing wastes are usually utilized via cattle feeding, burning and landfills, but they contain significant amounts of carbohydrates, proteins and lipids that could be used to produce valuable compounds such as lactic acid. Research with different agricultural (orange, banana and potato) peel wastes fermented by mixed cultures showed that lactic acid was the predominant chemical produced in all fermentation broths. The abundance of LAB rapidly increased during fermentation and genus *Lactobacillus* dominated at the end of process [139]. LAB, mainly *Lactobacillus* species, successfully produced lactic acid from other lignocellulosic substrates (**Table 1**).

Lignocellulosic substrates	Bacteria	Lactic acid production	References
Alfalfa fiber	*Lactobacillus delbrueckii*	0.606 g/g	[140]
	Lactobacillus pentosus	0.59 g/g	
Apple pomace	*Lactobacillus rhamnosus*	32.5 g/L	[141]
Cellulosic biosludges	*Lactobacillus rhamnosus*	39.4 g/L	[142]
Chips of oak wood	*Enterococcus faecalis*	24-93 g/L	[143]
Milled newspaper	*Lactobacillus delbrueckii*	24 g/L	[144]
Municipal solid waste	*Lactobacillus pentosus*	65 g/L	[145]
Pine needles	Co-culture of *Lactobacillus delbrueckii* and *Lactobacillus pentosus*	45.10 g/L	[146]
Recycled paper sludge	*Lactobacillus rhamnosus*	73 g/L	[147]
Sugarcane bagasse	*Lactococcus lactis*	10. 85 g/L	[148]
Turmeric residue	*Lactobacillus paracasei*	97.13 g/L	[149]
Vine-trimming wastes	*Lactobacillus pentosus*	21.8 g/L	[150]
Waste cardboard	*Lactobacillus coryniformis*	0.514 g/g	[151]
Wheat straw	*Lactobacillus pentosus*	6.6–6.7 g/L	[152]
	Lactobacillus brevis	4–4.7 g/L	
Wood chips of *Eucalyptus globulus*	*Lactobacillus delbrueckii*	48–62 g/L	[153]

Table 1. Lignocellulosic substrates in lactic acid production.

7. Use of LAB in production of bioenergy and biogas

The latest decades have witnessed growing interest in production of green energy. Fossil fuels adversely influence the environment owing to emission of carbon dioxide, triggering search for inexpensive renewable sources of energy that do not affect the surrounding nature. Microbial fuel cells (MFC) are devices that utilize organic and inorganic wastes and transform their chemical energy into electrical energy. MFC consist of anode and cathode chambers, physically separated by a proton exchange membrane (PEM). Microorganism in the anode section oxidizes the organic substrates and produces electrons and protons. The protons are conducted to the cathode chamber through PEM, and the electrons are conveyed via external circuit. Protons and electrons are reacting in the cathode chamber along with parallel reduction of oxygen to water. A steady current is generated by this process within the wire connecting anode and cathode. Besides generation of bioelectricity, MFC additionally resolve problem concerning utilization of waste [154].

MFC research has been conducted during several decades, but studies engaging LAB for generation of bioenergy were initiated only in recent years. Fe(III)-reducing bacterium

Enterococcus gallinarum MG25 turned out to be electrochemically active strain. It appears that MG25 can transfer electrons to the electrode as electron acceptor, so that the strain is expected to have promising MFC application prospects [155]. *L. lactis* is normally homolactic bacterium under anaerobic conditions. It lacks the genes that encode biosynthesis of heme. When a source of heme is provided, the respiratory chain is activated and the bacterium can oxidize NADH using O_2 as terminal electron acceptor. If lower potential terminal electron acceptors are engaged, such as hexacyanoferrate, ferric citrate or cupric chloride, the electron transfer chain is not required in its entirety up to cytochrome oxidase step, with final electron transmission carried out mostly by quinones. *L. lactis* was observed to perform extracellular electron transfer to anodes by utilizing at least two soluble redox mediators (one of these two mediators was 2-amino-3-dicarboxy-1,4-naphthoquinone) with acetate and pyruvate production and electricity generation [156]. Mixed cultures also can be used in MFC. While *Shewanella oneidensis* or *L. lactis* alone cannot generate electric current from glucose, they can do so in co-culture. *L. lactis* converts glucose into lactate, which serves as electron donor to *S. oneidensis* [157]. *Lactobacillus bulgaricus* was tested as producer of electricity. The maximum power (201.8 mW/m^2) was generated at optical density 0.5 by connecting in series MFC reactors with potassium permanganate as the electrolyte solution [158]. Further on, electricity output reached power density 393.23 mW/m^2 with LAB application [159]. Indium tin oxide (ITO) conductive glass anode modified by chitosan (CS) and α-Fe_2O_3 nanoparticles using LAB as the source of electrons raised considerably electricity generation. The maximum power density values of ITO blank, ITO/(CS/α-Fe_2O_3)$_4$/CS and ITO/(CS/α-Fe_2O_3)$_8$/CS were 0.035, 0.124 and 0.084W/m^2, respectively. The higher roughness of ITO/(CS/α-Fe_2O_3)$_4$/CS resulted in higher specific surface area available for growth of bacteria [160]. Following the trend, further development of MFC engaging LAB can be expected. Noteworthy, wastes are often applied in this technology, resolving thereby waste utilization problem.

Microbial electrolysis cell (MEC) is a technology similar to MFC, but this system recovers energy from substrates as valuable chemical compounds, like hydrogen. The latter is formed by reduction of protons with the transferred electrons in MEC. A microbial consortium demonstrated the ability to consume cheese whey as the sole carbon source yielding electricity or hydrogen. Cheese whey was fermented mainly by lactic acid bacteria (*Enterococcus* genus) and exoelectrogenic activity was expressed by *Geobacter* sp., utilizing acetate derived from fermentation as electron donor. The coulombic efficiency was 49±8% in the MFC system. In the MEC, hydrogen production reached 0.8 $L_{H2}/L_{REACTOR}$/d and it proved the potentiality of cheese whey to be a good carbon source for bioenergy production [161].

Added to MEC and MFC, LAB may be involved in the production of biofuels such as hydrogen, methane (biogas), ethanol and butanol. Hydrogen is one of the most attractive energy carriers alternative to conventional fossil fuels. It does not affect environment and produces only water vapor and heat energy as the result of its burning. Hydrogen is a highly efficient energy source; its specific energy value equals 33 Wh/g. For comparison, the specific energy of methane is 14.2 Wh/g and coal is 9.1 Wh/g. The biological processes leading to hydrogen production are dark fermentation, photofermentation, direct and indirect biophotolysis, as well as anaerobic respiration of sulfate-reducing bacteria under conditions of sulfate depletion [162]. LAB are unable to produce hydrogen themselves, but can influence hydrogen generation by increasing or decreasing its production. LAB can act as seeds for self-flocculated granule formation in hydrogen generation [163]. Another research revealed relation between the number of LAB and hydrogen production from simulated cheese processing wastewater via anaerobic fermentation using mixed microbial communities. More than 50% of the bacteria were *Lactobacillus* and about 5% of the isolates were hydrogen-producing *Clostridia* species. When H_2 production in the bioreactors decreased, concurrent reduction in the cell titer of genus *Lactobacillus* was also observed. It can be connected with pH important for H_2 production [164]. *Leuconostocaeae* were one of the predominant microbes in hydrogen-producing consortia in other experiment. Their role in the process is discussed [165]. When mixed cultures were used, *Lactobacillus amylovorus* utilized algal starch for lactic acid production and *Rhodobium marinum* produced hydrogen in the presence of light using lactic acid as an electron donor [166]. Products of LAB such as lactic acid also showed positive effect on hydrogen production. The addition of lactic acid to starch-containing medium could improve the hydrogen production rate and hydrogen production yield from 4.31 to 8.23 mL/h and from 5.70 to 9.08 mmol H_2/g starch, respectively. The authors guessed that enhanced hydrogen production was associated with a shift from acetic acid and ethanol formation to synthesis of butyric acid as the predominant metabolite. The increase in hydrogen yield was attributed to the increase in the available residual NADH for H_2 production. However, when lactic acid was used as the sole carbon source, no significant hydrogen generation was observed [167]. *Clostridium diolis* JPCC H-3 on medium with acetic acid and lactic acid produced 2.85 mL H_2/5 mL solution as compared to the control (0.63/5 mL solution) [168]. *Rhodobacter sphaeroides* GL-1 immobilized on polyurethane foam in a continuous flow bioreactor converted lactic acid to H_2 with an efficiency of 86% [169]. The hydrogen yield of *R. sphaeroides* RV was found to depend on lactic acid concentration, and maximum bacterial activity was observed at 100 mM influent lactic acid [170]. Nevertheless, other studies showed negative influence of LAB on hydrogen production. *L. paracasei*, *Enterococcus durans* and their supernatants inhibited hydrogen production via excretion of bacteriocins which have a deleterious effect on other bacteria. The inhibition of hydrogen production can be reduced by heat treatment for 30 min at temperatures ranging from 50 to 90°C and partially removed in the presence of protease trypsin inactivating bacteriocins [171]. The bacteriocin-producing LAB (mostly *Lactobacillus* spp.) were found to suppress hydrogen production during fermentation of cheese whey wastewater. At the same time, the highest H_2 yields were obtained when growth of *Lactococcus* spp. was associated with *Leuconostoc pseudomesenteroides*, although *Lactococcus* spp. is not recognized as hydrogen-producing strain [172]. Competition for resources between bacteria also reduces hydrogen production [173–175].

Biogas is a renewable energy source, which can be used as gaseous vehicle fuel and replace natural gas as a feedstock for producing chemicals and materials. Concerning biogas production, LAB are not directly involved in its generation, but the bacteria are able to influence methane yield. Crop characteristics, process parameters and management measures have a major impact on biogas yield. Ensiling with inoculated LAB is an appropriate method of storing feedstock for biogas production. Ensiling, prolonged storage and biological silage additives showed positive effects on methane yield of up to 11%. These could be attributed to increase in ratio of organic acids and alcohols. Changes in composition of fermentation products during ensiling and storage duration compensate for silage losses. Silage additives either accelerate the ensiling process or stabilize the silage [176, 177]. Different crops showed various need in ensiling promoters. Additive-free ensiling resulted in minor losses (0–13%) in the methane potential of sugar beet tops, but more substantial losses (17–39%) in the methane potential of grass. Ensiling with supplements improved the methane potential of both substrates by 19–22% [178]. High concentrations of ethanol and butyric acid following clostridial and heterofermentative lactic acid bacterial fermentations were also accompanied by elevated specific CH_4 yield from grass [179]. The methane yield of maize silage treated with heterofermentative LAB was measured higher than from the corresponding solid residue, while the treatment of amaranth showed a significant decrease in methane yield from silage in contrast to solid residue [180]. Other studies showed that LAB failed to raise methane yield or had little effect [181–183]. One experiment indicated that silage from maize straw not exposed to ensiling preparations was characterized by the highest biogas yield [184]. LAB could not stimulate total methane production, but they were able to promote the methane production rate at the beginning of the process [183]. The other products, like food waste, could also serve as methane sources. However, lactic acid pre-fermentation of food waste caused acid inhibition of the methanogenesis. Methane yield was a bit higher compared to the control, but significantly lower when ethanol pre-fermentation was used [185]. Lactic acid exerted extremely negative influence on methanogenesis of kitchen waste [186]. Although application of LAB in the ensiling process does not always increase methane yield, these bacteria conduce preservation of silage used in biogas production. LAB lead to PH drop by producing organic acids (mainly lactic and acetic acids) and decrease risk of microbial contamination [187].

Ethanol is another renewable energy source derived from plant biomass. Global production of ethanol increased from 17.25 billion L in 2000 to over 46 billion L in 2007 [188]. Yeasts are one of the main producers of ethanol. Nevertheless, ethanol generation process may be influenced by several factors, including microbial contamination. LAB are very abundant in the process because of their tolerance to ethanol, low pH and high temperature. Some strains are able to grow in media with 16% ethanol [189]. Diverse species of LAB can be found in the bioethanol process [190, 191]. It was shown that lactic acid may affect yeast viability [192]. However, due to the above-mentioned features and ability to produce ethanol (heterofermentative pathway), LAB can also be considered as biofuel sources. *L. buchneri* NRRL B-30929 ferments solely glucose at pH 4.0 into lactate and ethanol at molar ratio 1.03:1. Equimolar amounts of ethanol and lactate are produced when only xylose is available for the strain [193]. Recombinant strain *L. plantarum* containing several genes of *Sarcina ventriculi* produced slightly more ethanol (90–130 mM) than the control [194].

Biobutanol is another promising fuel. Compared to ethanol, butanol is distinguished by higher energy content, higher octane number, lower latent heat, lower solubility in water, higher vapor pressure and inferior corrosive capacity. Additionally, butanol can be directly included in the current design of internal combustion engines. The species *Clostridia* are the natural producers of butanol. However, they are difficult to culture and butanol is characterized by toxicity to bacteria at concentrations over 20 g/L, far below its solubility in water (~70 g/L) [195]. As a consequence, other microorganisms are screened for butanol production. Due to high degree of alcohol tolerance, LAB became objects for genetic manipulations to select butanol-producing strains. The recombinant *L. brevis* strain containing the clostridial genes *crt, bcd, etfB, etfA* and *hbd* was able to synthesize up to 300 mg/L butanol comparable to recombinant *E. coli* (580 and 552 mg/L) and

Pseudomonas putida (120 mg/L) cultures [196–198]. Recombinant strains of *L. lactis* and *L. buchneri* containing clostridial thiolase produced about 28 and 66 mg/L butanol, respectively [199]. Some *L. brevis* strains were found to produce 2-butanol without recombination. These strains converted meso-2,3-butanediol to 2-butanol in a synthetic medium, but none of them showed the same ability in a complex medium such as MRS. It appears that the process is inhibited by some kind of repression mechanism [200].

LAB effects on energy generation are controversial. It was shown that LAB can be used for energy generation in MFC and MEC and production of butanol. However, influence of the bacteria on hydrogen, biogas and ethanol processes is complicated. LAB fail to generate hydrogen and biogas, but they and their products are able to increase or decrease the output of biofuels. Concerning ethanol, LAB may reduce yeast product yields or act as substrate providers. Contradictory impact of LAB on bioenergy generation requires further research to minimize negative effects and gain maximum benefits.

8. Use of LAB in food industry

Fermentation is the important process for manufacturing of products with desirable biochemical characteristics with the aid of microorganisms or enzymes. Fermentation plays at least five roles:

1. Enrichment of the diet through development of a diversity of flavors, aromas and textures in food substrates.
2. Preservation of food via lactic acid, ethanol, acetic acid and alkaline fermentations.
3. Biological upgrading of food substrates with proteins, essential amino acids, fatty acids and vitamins.
4. Detoxification in the course of food fermentation.
5. Saving cooking time and fuel requirements [201].

LAB from ancient times have been used in production of traditional foodstuffs. LAB are important microorganisms involved in manufacturing various dairy products such as yogurt, kefir, cheese, butter and so on. The latter account for about 20% of the global output of fermented products [202]. LAB can be divided into two groups depending on optimal growth temperature: mesophilic (20–30°C) and thermophilic (30–45°C). The flavor, texture and consistency may vary considerably when mesophilic or thermophilic cultures are used. Dairy industry mainly consumes starter cultures selected and maintained by subcultivation in milk. Several steps are carried out to obtain the required products [203, 204]:

1. Selection of starter cultures, optimization of medium and cultural conditions. These factors affect the yield of the product and its characteristics.

2. Pretreatment. This step includes various processes such as clarification, fat separation, standardization, evaporation, de-aeration, homogenization, pasteurization and so on. Pretreatment aim is to adjust dairy substrate characteristics and eliminate microorganisms able to interfere with fermentation process. The milk is then cooled to the appropriate fermentation temperature.

3. Fermentation. After pretreatment step, starter cultures are added and incubated at optimal temperature for the definite period. The bacteria ingest the lactose and release some compounds, like lactic acid. Production of lactic acid results in increased acidity causing milk proteins to denature and aggregate and growth inhibition of other acid-sensitive species.

4. Postfermentation step. After the end of fermentation process, the product may be subjected to downstream processing and upgrading (addition of flavorings, homogenization, filtration, etc.).

5. Packing, labeling, storage and market distribution of the product.

Manufacturing of fermented meat, soy, vegetables and baking products using LAB is carried out by a similar scheme. LAB provide the characteristic flavor and produce acids (e.g. lactic acid) that lower pH of the products and inhibit growth of spoilage microorganisms [205].

As mentioned in Section 5, LAB are sources of various compounds that can be used as food additives. Studies showed high efficiency of LAB in product enrichment with these additives. *L. amylovorus* CRL887 was able to produce significant concentrations of folate, or vitamin B_9 (81.2 ± 5.4 μg/L), on folate-free cultural medium. Co-fermentation with B9 producing starter cultures *L. bulgaricus* CRL871 and *Streptococcus thermophilus* CRL803 and CRL415 yielded yogurt with high folate concentration (263.1 ± 2.4 μg/L). A single portion of the product provides for 15% of the recommended dietary allowance [206]. *L. plantarum* was shown to increase about twofold and threefold riboflavin (vitamin B_2) content in pasta and bread, respectively [207]. *L. reuteri* CRL1098 from sourdough was able to produce vitamin B_{12} or cobalamin [208]. *L. lactis* ssp. *cremoris* YIT 2012 and *Leuconostoc lactis* YIT 3001 produced 9–123 μg of vitamin K2/L in defatted dry milk and soymilk medium, respectively, providing beneficial property for dietary supplement [209].

Concerning bacteriocins, nisin has been approved worldwide to use as a natural food preservative in food industry. It demonstrated a long record of food preservation efficiency [210]. Other bacteriocins also have practical applications. Paracin C produced by *L. paracasei* CICC 20241 induced extensive cell damage and disintegration of *Alicyclobacillus acidoterrestris* causing spoilage of fruit juices. The bacteriocin additionally reduced thermal resistance of bacterial spores [211]. *L. paracasei* subsp. *tolerans* from kefir produced bacteriocin inhibiting both fungi and bacteria [212]. Bacteriocin produced by *P. acidilactici* showed suppressing and bactericidal effect on *L. monocytogenes* in meat products [213]. *Lactobacillus* species isolated from different fermented cereal gruels demonstrated inhibitory action on growth of various target organisms [214]. Bacteriocin of *Enterococcus faecium* CN-25 isolated from fermented fish product completely inhibited growth of *L. monocytogenes* at the minimum concentration 2.38 mg/mL [215].

CLA-producing strains may be used in the food industry to derive products with increased CLA content. Strains of the genera *Bifidobacterium*, *Lactobacillus* and *Lactococcus* are able to enrich skim milk with CLA (in the range of 40–50 μg CLA/mL) [216]. Administration of *Lactobacillus* strains led to significant increase in CLA concentrations 0.2–1.2 mg/g fat in eggs and 0.3–1.88 mg/g fat in broiler chicken cuts [217]. *L. plantarum* from fermented pickle brines exhibited high CLA-producing ability in the presence of linoleic acid [218].

EPS of LAB have a wide application range. They can be used to modify certain food features. Incorporation of EPS may provide viscosity, stability and water-binding functions that may contribute positively to the odor, texture and taste of fermented dairy products [219]. *S. thermophilus* zlw TM11 induced high exopolysaccharide content (380 mg/L) and viscosity (7716 mPa/s) of fermented milk. The co-culture of this strain with *L. delbrueckii* subsp. *bulgaricus* 3 4.5 caused low syneresis (8.5%), better texture and sensory perception of fermented yogurt [220]. EPS from *S. thermophilus* MR-1C significantly increased moisture retention in low-fat mozzarella. The cheese with low moisture content has a tough and rubbery texture and requires more heat for melting [221]. EPS-producing LAB were used in the production of Swedish ropy milk with proper level of viscosity [222]. Sour cream fermented by *S. thermophilus* strains producing capsular exopolysaccharides was characterized by low syneresis, high apparent viscosity and increased adhesiveness and gumminess [223]. EPS-producing strains of *S. thermophilus* showed reduced freezing mortality when LAB were introduced into frozen dairy desserts as a source of β-galactosidase hydrolyzing lactose and producing the absorbable monosaccharides glucose and galactose [224]. Besides dairy industry, EPS are used in bakery. *Weissella cibaria* WC4 and *L. plantarum* LP9 were able to produce EPS that increased the viscosity of baked product and the resulting bread was distinguished by higher specific volume and lower firmness [225]. EPS can improve not only taste, structure, consistency and shelf life of food products but also probiotic characteristics. Fermented milk with EPS-producing *S. thermophilus* culture and purified EPS resuspended in milk were effective for gastritis prevention [226]. Three strains of *L. delbrueckii* subsp. *bulgaricus* isolated from home-made yogurt produced high amounts of EPS and showed cholesterol lowering effects [227].

Reactive oxygen species and free radicals take part in the development of degenerative diseases such as cancer, atherosclerosis and diabetes [228]. Foods containing antioxidative materials may be applied for prevention of these diseases. LAB demonstrated antioxidant activity and could be used in the production of food with required properties. The radical-scavenging activity of water/salt-soluble extracts from sourdough fermented by pool of LAB was significantly higher than in control chemically acidified dough. The highest activity was found for whole wheat, spelt, rye and kamut sourdoughs [85]. It was also demonstrated that LAB strains were able to produce antioxidant activity in dairy products. The formation of 4–20 kDa peptides was accompanied by elevated radical scavenging activity [229]. *L. plantarum* KFRI 00144, *L. delbrueckii* subsp. *latis* KFRI 01181, *Bifidobacterium breve* KFRI K-101 and *Bifidobacterium thermophilum* KFRI 00748 were able to efficiently biotransform isoflavone glucosides to their bioactive aglycones during soybean fermentation. Isoflavones are known for their potential bioactive antioxidant properties and radical scavenging capacity. It has been shown that isoflavone glucosides were poorly absorbed in the small intestine compared with their aglycones, so that soybean fermented by LAB could be regarded as a potent antioxidant and radical scavenging dietary source [230].

LAB are able to alter flavor and taste characteristics of fermented food. Prolonged wheat and rye fermentations performed by LAB resulted in sourdoughs with acidic (*Lactobacillus fermentum* IMDO 130101, *L. plantarum* IMDO 130201 and *Lactobacillus crustorum* LMG 23699), butter-like (*L. amylovorus* DCE 471), or fruity flavor (*L. sakei* CG1). Carbonyls, including alcohols, acids, aldehydes, hydrocarbon-substituted furans, ketones, esters, pyrazines and pyrrolines, are recognized as important bread flavoring agents [231]. Concerning cheese, proteolysis and the subsequent amino acid catabolism are of primary significance for the development of flavor, irrespective of the cheese variety. Amino acids are major precursors for volatile aroma compounds [94]. Taste and flavor of wines are determined by alcoholic and the following malolactic fermentation. Most red and white wines upon malolactic fermentation display more exquisite taste, with an improved bouquet. On the contrary, light red wines and some white wines are characterized by the grape aromas and the vivacity which fades with malolactic fermentation [26].

Production of polyols such as mannitol by bacterial fermentation is a promising method. Fermentation process could have several advantages over the chemical synthesis, such as complete conversion of fructose to mannitol, absence of hardly disposable side products, moderate production conditions and no strict need of highly purified substrates [232]. However, mannitol is still produced industrially by high pressure hydrogenation of fructose/glucose mixtures in aqueous solution at high temperature [233]. It is the same case with other polyols [234].

9. Conclusion

LAB represent a versatile group of microorganisms. Owing to their valuable properties, LAB have been used in food production since ancient times. Development of natural sciences led to discovery of LAB as normal part of human and animal microflora. LAB are recognized as safe microorganisms and they are mainly applied in food industry for production of dairy, meat, bread, fish and vegetable products and in medicine as probiotics. LAB are known to synthesize a wide range of compounds consumed in various areas. LAB produce bacteriocins, vitamins, low calorie sugars, EPS and other valuable substances regarded as additives improving safety, quality and flavor of foodstuffs. However, one of the main LAB products is lactic acid used in food processing, pharmaceutics, cosmetics and other industrial sectors. Steadily growing market demand for this commodity urges researchers and manufacturers to seek less expensive substrates for its synthesis. Many studies deal with industrial and household wastes as appropriate sources for lactic acid production.

Ongoing research revealed encouraging LAB application prospects in other fields, such as agriculture, bioremediation of environment, chemical industry and so on. Need in green energy instead of fossil fuels focused keen interest on bacteria as sources of energy, including

LAB. Despite contradictory results, further investigations could resolve problems caused by inhibitory effects of LAB and thus increase biofuel yields.

Author details

Galina Novik[1]*, Olga Meerovskaya[2] and Victoria Savich[1]

*Address all correspondence to: galina_novik@mbio.bas-net.by

1 National Academy of Sciences of Belarus, Institute of Microbiology, Minsk, Belarus

2 Belarusian Institute of System Analysis and Information Support for Scientific and Technical Sphere, Minsk, Belarus

References

[1] Fennema OF, Hui YH, Karel M, Walstra P, Whitaker JR. Lactic acid bacteria (microbiological and functional aspects). In: Salminen S, von Wright A, editors. Food Science and Technology a Series of Monographs, Textbooks, and Reference Books. 3rd ed. New York: Marcel Dekker, Inc.; 2004. pp. 19-30.

[2] Leroy F, De Vuyst L. Lactic acid bacteria as functional starter cultures for the food fermentation industry. Trends in Food Science & Technology. 2004;**15**(2):67-78. DOI: 10.1016/j.tifs.2003.09.004

[3] Orla-Jensen S. The Lactic Acid Bacteria. Copenhagen: Host and Son; 1919.

[4] Liu W, Pang H, Zhang H, Cai Y. Biodiversity of lactic acid bacteria. In: Zhang H, Cai Y, editors. Lactic Acid Bacteria. Netherlands: Springer; 2014. pp. 103-203. DOI: 10.1007/978-94-017-8841-0_2

[5] Poupard JA, Husain I, Norris RF. Biology of the bifidobacteria. Bacteriological Reviews. 1973;**37**(2):136-165.

[6] Temmerman R, Huys G, Swings J. Identification of lactic acid bacteria: culture-dependent and culture-independent methods. Trends in Food Science & Technology. 2004;**15**(7):348-359. DOI: 10.1016/j.tifs.2003.12.007

[7] Patel JB. 16S rRNA gene sequencing for bacterial pathogen identification in the clinical laboratory. Molecular Diagnosis. 2001;**6**(4):313-21. DOI: 10.1007/BF03262067

[8] Amor KB, Vaughan EE, de Vos WM. Advanced molecular tools for the identification of lactic acid bacteria. The Journal of Nutrition. 2007;**137**(3 Suppl 2):741S-747S.

[9] Stieglmeier M, Wirth R, Kminek G, Moissl-Eichinger C. Cultivation of anaerobic and facultatively anaerobic bacteria from spacecraft-associated clean rooms. Applied and Environmental Microbiology. 2009;**11**(75):3484-3491. DOI: 10.1128/AEM.02565-08.

[10] Lahtinen S, Ouwehand AC, Salminen S, von Wright A, editors. Lactic Acid Bacteria: Microbiological and Functional Aspects. 4th ed. USA: CRC Press; 2011. 798 p.

[11] Idler C, Venus J, Kamm B. Microorganisms for the production of lactic acid and organic lactates. In: Kamm B, editor. Microorganisms in Biorefineries. Berlin, Heidelberg: Springer; 2015. pp. 225-273. DOI: 10.1007/978-3-662-45209-7_9

[12] Kandler O. Carbohydrate metabolism in lactic acid bacteria. Antonie van Leeuwenhoek. 1983;**49**(3):209-224. DOI: 10.1007/BF00399499

[13] Liu M, Nauta A, Francke C, Siezen RJ. Comparative genomics of enzymes in flavor-forming pathways from amino acids in lactic acid bacteria. Applied and Environmental Microbiology. 2008;**74**(15):4590-4600. DOI: 10.1128/AEM.00150-08.

[14] Liu M, Bayjanov JR, Renckens B, Nauta A, Siezen RJ. The proteolytic system of lactic acid bacteria revisited: a genomic comparison. BMC Genomics. 2010;**11**:36. DOI: 10.1186/1471-2164-11-36.

[15] Meyers SA, Cuppett SL, Hutkins RW. Lipase production by lactic acid bacteria and activity on butter oil. Food Microbiology. 1996;**13**(5):383-389. DOI: 10.1006/fmic.1996.0044

[16] Ogawa J, Kishino S, Ando A, Sugimoto S, Mihara K, Shimizu S. Production of conjugated fatty acids by lactic acid bacteria. Journal of Bioscience and Bioengineering. 2005;**100**(4):355-364. DOI: 10.1263/jbb.100.355

[17] Holland R, Liu SQ, Crow VL, Delabre ML, Lubbers M, Bennett M, Norris G. Esterases of lactic acid bacteria and cheese flavour: Milk fat hydrolysis, alcoholysis and esterification. International Dairy Journal. 2005;**15**(6):711-718. DOI: 10.1016/j.idairyj.2004.09.012

[18] Chelule PK, Mokoena MP, Gqaleni N. Advantages of traditional lactic acid bacteria fermentation of food in Africa. In: Méndez-Vilas A, editor. Current research, technology and education topics in applied microbiology and microbial biotechnology. Vol. 2. Spain: Formatex; 2010. pp. 1160-1167.

[19] Rhee SJ, Lee JE, Lee CH. Importance of lactic acid bacteria in Asian fermented foods. Microbial Cell Factories. 2011;**10**(Suppl 1):S5. DOI: 10.1186/1475-2859-10-S1-S5

[20] Kongo JM. Lactic acid bacteria – R & D for food, health and livestock purposes. Rijeka, Croatia: InTech; 2013. 658 p. DOI: 10.5772/2825

[21] Gemechu T. Review on lactic acid bacteria function in milk fermentation and preservation. African Journal of Food Science. 2015;**9**(4):170-175. DOI: 10.5897/AJFS2015.1276

[22] El-Ghaish S, Ahmadova A, Hadji-Sfaxi I, El Mecherfi KE, Bazukyan I, Choiset Y, Rabesona H, Sitohy M, Popov YG, Kuliev AA, Chobert J-M, Haertle T, Mozzi F. Potential use of lactic acid bacteria for reduction of allergenicity and for longer conservation of fermented foods. Trends in Food Science & Technology. 2011;**22**(9):509-516. DOI: 10.1016/j.tifs.2011.05.003

[23] Arendt EK, Moroni A, Zannini E. Medical nutrition therapy: use of sourdough lactic acid bacteria as a cell factory for delivering functional biomolecules and food ingredients in gluten free bread. Microbial Cell Factories. 2011;**10**(Suppl 1):S15. DOI: 10.1186/1475-2859-10-S1-S15

[24] Glatman L, Drabkin V, Gelman A. Using lactic acid bacteria for developing novel fish food products. Journal of the Science of Food and Agriculture. 2000;**80**(3):375-380. DOI: 10.1002/1097-0010(200002)80:3<375::AID-JSFA539>3.0.CO;2-S

[25] Françoise L. Occurrence and role of lactic acid bacteria in seafood products. Food Microbiology. 2010;**27**(6):698-709. DOI: 10.1016/j.fm.2010.05.016

[26] Lonvaud-Funel A. Lactic acid bacteria in the quality improvement and depreciation of wine. Antonie Van Leeuwenhoek. 1999;**76**(1-4):317-331. DOI: 10.1023/A:1002088931106

[27] Naidu AS, Bidlack WR, Clemens RA. Probiotic spectra of lactic acid bacteria (LAB). Critical Reviews in Food Science and Nutrition. 1999;**39**(1):13-126. DOI: 10.1080/10408699991279187

[28] Reis JA, Paula AT, Casarotti SN, Penna ALB. Lactic acid bacteria antimicrobial compounds: Characteristics and applications. Food Engineering Reviews. 2012;**4**(2):124-140. DOI: 10.1007/s12393-012-9051-2

[29] Fioramonti J, Theodorou V, Bueno L. Probiotics: What are they? What are their effects on gut physiology? Best Practice & Research Clinical Gastroenterology. 2003;**17**(5):711-724. DOI: 10.1016/S1521-6918(03)00075-1

[30] Heyman M. Effect of lactic acid bacteria on diarrheal diseases. Journal of the American College of Nutrition. 2000;**19**(2 Suppl):137S-146S. DOI: 10.1080/07315724.2000.10718084

[31] Wang MF, Lin HC, Wang YY, Hsu CH. Treatment of perennial allergic rhinitis with lactic acid bacteria. Pediatric Allergy and Immunology. 2004;**15**(2):152-158. DOI: 10.1111/j.1399-3038.2004.00156.x

[32] Pohjavuori E, Viljanen M, Korpela R, Kuitunen M, Tiittanen M, Vaarala O, Savilahti E. *Lactobacillus* GG effect in increasing IFN-gamma production in infants with cow's milk allergy. The Journal of Allergy and Clinical Immunology. 2004;**114**(1):131-136. DOI: 10.1016/j.jaci.2004.03.036

[33] Hayashi A, Kimura M, Nakamura Y, Yasui H. Anti-atopic dermatitis effects and the mechanism of lactic acid bacteria isolated from Mongolian fermented milk. The Journal of Dairy Research. 2009;**76**(2):158-164. DOI: 10.1017/S0022029908003725.

[34] del Carmen S, de Moreno de LeBlanc A, Miyoshi A, Santos Rocha C, Azevedo V, LeBlanc JG. Potential application of probiotics in the prevention and treatment of inflammatory bowel diseases. Ulcers. 2011;**2011**:1-13. DOI: 10.1155/2011/841651

[35] Solga SF. Probiotics can treat hepatic encephalopathy. Medical Hypotheses. 2003;**61**(2):307-313. DOI: 10.1016/S0306-9877(03)00192-0

[36] Zhong L, Zhang X, Covasa M. Emerging roles of lactic acid bacteria in protection against colorectal cancer. World Journal of Gastroenterology. 2014;**20**(24):7878-7886. DOI: 10.3748/wjg.v20.i24.7878

[37] Alm L. Effect of fermentation on lactose, glucose, and galactose content in milk and suitability of fermented milk products for lactose intolerant individuals. Journal of Dairy Science. 1982;**65**(3):346-352. DOI: 10.3168/jds.S0022-0302(82)82198-X

[38] Agerholm-Larsen L, Bell ML, Grunwald GK, Astrup A. The effect of a probiotic milk product on plasma cholesterol: A meta-analysis of short-term intervention studies. European Journal of Clinical Nutrition. 2000;**54**(11):856-860.

[39] Hlivak P, Odraska J, Ferencik M, Ebringer L, Jahnova E, Mikes Z. One-year application of probiotic strain *Enterococcus faecium* M-74 decreases serum cholesterol levels. Bratislavské lekárske listy. 2005;**106**(2):67-72.

[40] Tsai JS, Lin YS, Pan BS, Chen TJ. Antihypertensive peptides and γ-aminobutyric acid from prozyme 6 facilitated lactic acid bacteria fermentation of soymilk. Process Biochemistry. 2006;**41**(6):1282-1288. DOI: 10.1016/j.procbio.2005.12.026

[41] Shida K, Kiyoshima-Shibata J, Kaji R, Nagaoka M, Nanno M. Peptidoglycan from lactobacilli inhibits interleukin-12 production by macrophages induced by *Lactobacillus case*i through Toll-like receptor 2-dependent and independent mechanisms. Immunology. 2009;**128**(1 Suppl):e858-869. DOI: 10.1111/j.1365-2567.2009.03095.x.

[42] Di Cagno R, Mazzacane F, Rizzello CG, Vincentini O, Silano M, Giuliani G, De Angelis M, Gobbetti M. Synthesis of isoflavone aglycones and equol in soy milks fermented by food-related lactic acid bacteria and their effect on human intestinal Caco-2 cells. Journal of Agricultural and Food Chemistry. 2010;**58**(19):10338-10346. DOI:10.1021/jf101513r

[43] Pophaly SD, Singh P, Kumar H, Tomar SK, Singh R. Selenium enrichment of lactic acid bacteria and bifidobacteria: A functional food perspective. Trends in Food Science & Technology. 2014;**39**(2):135-145. DOI: 10.1016/j.tifs.2014.07.006

[44] Nácher-Vázquez M, Ballesteros N, Canales Á, Rodríguez Saint-Jean S, Pérez-Prieto SI, Prieto A, Aznar R, López P. Dextrans produced by lactic acid bacteria exhibit antiviral and immunomodulatory activity against salmonid viruses. Carbohydrate Polymers. 2015;**124**:292-301. DOI: 10.1016/j.carbpol.2015.02.020.

[45] Franz CM, Huch M, Abriouel H, Holzapfel W, Gálvez A. Enterococci as probiotics and their implications in food safety. International Journal of Food Microbiology. 2011;**151**(2):125-140. DOI: 10.1016/j.ijfoodmicro.2011.08.014

[46] Gaggìa F, Mattarelli P, Biavati B. Probiotics and prebiotics in animal feeding for safe food production. International Journal of Food Microbiology. 2010;**141**:S15-S28. DOI: 10.1016/j.ijfoodmicro.2010.02.031

[47] Zain ME. Impact of mycotoxins on humans and animals. Journal of Saudi Chemical Society. 2011;**15**(2):129-144. DOI: 10.1016/j.jscs.2010.06.006

[48] Broberg A, Jacobsson K, Ström K, Schnürer J. Metabolite profiles of lactic acid bacteria in grass silage. Applied and Environmental Microbiology. 2007;**73**(17):5547-5552. DOI: 10.1128/AEM.02939-06

[49] Franco TS, Garcia S, Hirooka EY, Ono YS, dos Santos JS. Lactic acid bacteria in the inhibition of *Fusarium graminearum* and deoxynivalenol detoxification. Journal of Applied Microbiology. 2011;**111**(3):739-748. DOI: 10.1111/j.1365-2672.2011.05074.x.

[50] Dalié DKD, Deschamps AM, Richard-Forget F. Lactic acid bacteria – Potential for control of mould growth and mycotoxins: A review. Food Control. 2010;**21**(4):370-380. DOI: 10.1016/j.foodcont.2009.07.011

[51] Bovo F, Corassin CH, Rosim RE, de Oliveira CA. Efficiency of lactic acid bacteria strains for decontamination of aflatoxin M1 in phosphate buffer saline solution and in skimmed milk. Food and Bioprocess Technology. 2013;**6**(8):2230-2234. DOI: 10.1007/s11947-011-0770-9

[52] Magnusson J, Ström K, Roos S, Sjögren J, Schnürer J. Broad and complex antifungal activity among environmental isolates of lactic acid bacteria. FEMS Microbiology Letters. 2003;**219**(1):129-135. DOI: 10.1016/S0378-1097(02)01207-7

[53] Varsha KK, Priya S, Devendra L, Nampoothiri KM. Control of spoilage fungi by protective lactic acid bacteria displaying probiotic properties. Applied Biochemistry and Biotechnology. 2014;**172**(7):3402-3413. DOI: 10.1007/s12010-014-0779-4.

[54] Crowley S, Mahony J, van Sinderen D. Current perspectives on antifungal lactic acid bacteria as natural bio-preservatives. Trends in Food Science & Technology. 2013;**33**(2):93-109. DOI: 10.1016/j.tifs.2013.07.004.

[55] O'Callahan DR, Singh T, McDonald IR. Evaluation of lactic acid bacterium from chilli waste as a potential antifungal agent for wood products. Journal of Applied Microbiology. 2012;**112**(3):436-442. DOI: 10.1111/j.1365-2672.2011.05226.x

[56] Pflügl S, Marx H, Mattanovich D, Sauer M. 1,3-Propanediol production from glycerol with *Lactobacillus diolivorans*. Bioresource Technology. 2012;**119**:133-140. DOI: 10.1016/j.biortech.2012.05.121

[57] Sharma D, Saharan BS, Kapil S. Biosurfactants of Lactic Acid Bacteria. Switzerland: Springer International Publishing; 2016. 86 p. DOI: 10.1007/978-3-319-26215-4

[58] Elbanna K, Hassan G, Khider M, Mandour R. Safe biodegradation of textile azo dyes by newly isolated lactic acid bacteria and detection of plasmids associated with degradation. Journal of Bioremediation & Biodegradation. 2010;**1**:110-118. DOI: 10.4172/2155-6199.1000110

[59] Coleman JK, Brown MJ, Moses V, Burton CC. Enhanced Oil Recovery. WO1992015771 A1. September 17, 1992

[60] Gustavsson J, Cederberg C, Sonesson U, Van Otterdijk R, Meybeck A. Global food losses and food waste – Extent, causes and prevention. Rome: Food and Agriculture Organization (FAO); 2011. 29 p.

[61] Kiran EU, Trzcinski AP, Ng WJ, Liu Y. Bioconversion of food waste to energy: A review. Fuel. 2014;**134**:389-399. DOI: 10.1016/j.fuel.2014.05.074

[62] Vázquez JA, Murado MA. Enzymatic hydrolysates from food wastewater as a source of peptones for lactic acid bacteria productions. Enzyme and Microbial Technology. 2008;**43**(1):66-72. DOI: 10.1016/j.enzmictec.2008.01.015

[63] Vázquez JA, Nogueira M, Durán A, Prieto MA, Rodríguez-Amado I, Rial D, González MP, Murado MA. Preparation of marine silage of swordfish, ray and shark visceral waste by lactic acid bacteria. Journal of Food Engineering. 2011;**103**(4):442-448. DOI: 10.1016/j.jfoodeng.2010.11.014

[64] Thomsen MH, Kiel P. Selection of lactic acid bacteria for acidification of brown juice (grass juice), with the aim of making a durable substrate for L-lysine fermentation. Journal of the Science of Food and Agriculture. 2008;**88**(6):976-983. DOI: 10.1002/jsfa.3176

[65] Deshmukh AC, Patterson PH. Preservation of hatchery waste by lactic acid fermentation. 1. Laboratory scale fermentation. Poultry Science. 1997;**76**(9):1212-1219.

[66] Deshmukh AC, Patterson PH. Preservation of hatchery waste by lactic acid fermentation. 2. Large-scale fermentation and feeding trial to evaluate feeding value. Poultry Science. 1997;**76**(9):1220-1226.

[67] Cleveland J, Montville TJ, Nes IF, Chikindas ML. Bacteriocins: safe, natural antimicrobials for food preservation. International Journal of Food Microbiology. 2001;**71**(1):1-20. DOI: 10.1016/S0168-1605(01)00560-8

[68] Zacharof MP, Lovitt RW. Bacteriocins produced by lactic acid bacteria a review article. APCBEE Procedia. 2012;**2**:50-56. DOI: 10.1016/j.apcbee.2012.06.010

[69] Yang SC, Lin CH, Sung CT, Fang JY. Antibacterial activities of bacteriocins: Application in foods and pharmaceuticals. Frontiers in Microbiology. 2014;**5**:241. DOI: 10.3389/fmicb.2014.00241

[70] Calo-Mata P, Arlindo S, Boehme K, de Miguel T, Pascoal A, Barros-Velazquez J. Current applications and future trends of lactic acid bacteria and their bacteriocins for the biopreservation of aquatic food products. Food and Bioprocess Technology. 2008;**1**(1):43-63. DOI 10.1007/s11947-007-0021-2

[71] Sobrino-López A, Martín-Belloso O. Use of nisin and other bacteriocins for preservation of dairy products. International Dairy Journal. 2008;**18**(4):329-343. DOI: 10.1016/j.idairyj.2007.11.009

[72] Settanni L, Corsetti A. Application of bacteriocins in vegetable food biopreservation. International Journal of Food Microbiology. 2008;**121**(2):123-138. DOI: 10.1016/j.ijfoodmicro.2007.09.001

[73] Woraprayote W, Malila Y, Sorapukdee S, Swetwiwathana A, Benjakul S, Visessanguan W. Bacteriocins from lactic acid bacteria and their applications in meat and meat products. Meat Science. 2016;**120**:118-132. DOI: 10.1016/j.meatsci.2016.04.004

[74] Rogers LA, Whittier ED. Limiting factors in lactic acid fermentation. Journal of Bacteriology. 1928;16:211-229.

[75] Patel AK, Michaud P, Singhania RR, Soccol CR, Pandey A. Polysaccharides from probiotics: new developments as food additives. Food Technology and Biotechnology. 2010;**48**(4):451-463.

[76] De Vuyst L, Degeest B. Heteropolysaccharides from lactic acid bacteria. FEMS Microbiology Reviews. 1999;**23**(2):153-177. DOI: 10.1111/j.1574-6976.1999.tb00395.x

[77] Patel S, Majumder A, Goyal A. Potentials of exopolysaccharides from lactic acid bacteria. Indian Journal of Microbiology. 2012;**52**(1):3-12. DOI: 10.1007/s12088-011-0148-8

[78] Saha BC, Nakamura LK. Production of mannitol and lactic acid by fermentation with *Lactobacillus intermedius* NRRL B-3693. Biotechnology and Bioengineering. 2003;**82**(7): 864-871. DOI: 10.1002/bit.10638

[79] Kim P. Current studies on biological tagatose production using L-arabinose isomerase: A review and future perspective. Applied Microbiology and Biotechnology. 2004;**65**(3): 243-249. DOI: 10.1007/s00253-004-1665-8

[80] Ladero V, Ramos A, Wiersma A, Goffin P, Schanck A, Kleerebezem M, Hugenholtz J, Smid EJ, Hols P. High-level production of the low-calorie sugar sorbitol by *Lactobacillus plantarum* through metabolic engineering. Applied and Environmental Microbiology. 2007;**73**(6):1864-1872. DOI: 10.1128/AEM.02304-06

[81] Nyyssölä A, Pihlajaniemi A, Palva A, von Weymarn N, Leisola M. Production of xylitol from D-xylose by recombinant *Lactococcus lactis*. Journal of Biotechnology. 2005;**118**(1):55-66. DOI: 10.1016/j.jbiotec.2005.03.014

[82] Veiga-da-Cunha M, Santos H, Van Schaftingen E. Pathway and regulation of erythritol formation in *Leuconostoc oenos*. Journal of Bacteriology. 1993;**175**(13):3941-3948.

[83] Soleymanzadeh N, Mirdamadi S, Kianirad M. Antioxidant activity of camel and bovine milk fermented by lactic acid bacteria isolated from traditional fermented camel milk (Chal). Dairy Science & Technology. 2016;**96**(4):443-457. DOI: 10.1007/s13594-016-0278-1

[84] Abubakr MA, Hassan Z, Imdakim MMA. Antioxidant activity of lactic acid bacteria (LAB) fermented skim milk as determined by 1,1-diphenyl-2-picrylhydrazyl (DPPH) and ferrous chelating activity (FCA). African Journal of Microbiology Research. 2012;**6**(34):6358-6364. DOI: 10.5897/AJMR12.702

[85] Coda R, Rizzello CG, Pinto D, Gobbetti M. Selected lactic acid bacteria synthesize antioxidant peptides during sourdough fermentation of cereal flours. Applied and Environmental Microbiology. 2012;**78**(4):1087-1096. DOI: 10.1128/AEM.06837-11

[86] O'Connor EB, Barrett E, Fitzgerald G, Hill C, Stanton C, Ross RP. Production of vitamins, exopolysaccharides and bacteriocins by probiotic bacteria. In: Tamime AY, editor. Probiotic Dairy Products. Oxford, UK: Blackwell Publishing; 2005. pp. 167-194. DOI: 10.1002/9780470995785.ch8

[87] Capozzi V, Russo P, Dueñas MT, López P, Spano G. Lactic acid bacteria producing B-group vitamins: a great potential for functional cereals products. Applied Microbiology and Biotechnology. 2012;**96**(6):1383-1394. DOI: 10.1007/s00253-012-4440-2

[88] Kuhl GC, De Dea Lindner J. Biohydrogenation of linoleic acid by lactic acid bacteria for the production of functional cultured dairy products: A review. Foods. 2016;**5**(1):13. DOI: 10.3390/foods5010013

[89] Rodríguez H, Curiel JA, Landete JM, de las Rivas B, López de Felipe F, Gómez-Cordovés C, Mancheño JM, Muñoz R. Food phenolics and lactic acid bacteria. International Journal of Food Microbiology. 2009;**132**(2-3):79-90. DOI: 10.1016/j.ijfoodmicro.2009.03.025.

[90] van Beek S, Priest FG. Decarboxylation of substituted cinnamic acids by lactic acid bacteria isolated during malt whisky fermentation. Applied and Environmental Microbiology. 2000;**66**(12):5322-5328. DOI: 10.1128/AEM.66.12.5322-5328.2000

[91] Curiel JA, Rodríguez H, Landete JM, de las Rivas B, Muñoz R. Ability of *Lactobacillus brevis* strains to degrade food phenolic acids. Food Chemistry. 2010;**120**(1):225-229. DOI: 10.1016/j.foodchem.2009.10.012

[92] Svensson L, Sekwati-Monang B, Lutz DL, Schieber A, Gänzle MG. Phenolic acids and flavonoids in nonfermented and fermented red sorghum (*Sorghum bicolor* (L.) Moench). Journal of Agricultural and Food Chemistry. 2010;**58**(16):9214-9220. DOI: 10.1021/jf101504v.

[93] González L, Sacristán N, Arenas R, Fresno JM, Tornadijo ME. Enzymatic activity of lactic acid bacteria (with antimicrobial properties) isolated from a traditional Spanish cheese. Food Microbiology. 2010;**27**(5):592-597. DOI: 10.1016/j.fm.2010.01.004

[94] Smit G, Smit BA, Engels WJ. Flavour formation by lactic acid bacteria and biochemical flavour profiling of cheese products. FEMS Microbiology Reviews. 2005;**29**(3):591-610. DOI: 10.1016/j.femsre.2005.04.002

[95] Krajewska B. Application of chitin- and chitosan-based materials for enzyme immobilizations: A review. Enzyme and Microbial Technology. 2004;**35**(2-3):126-139. DOI: 10.1016/j.enzmictec.2003.12.013

[96] Rinaudo M. Chitin and chitosan: properties and applications. Progress in Polymer Science. 2006;**31**(7):603-632. DOI: 10.1016/j.progpolymsci.2006.06.001

[97] Nishimura K, Nishimura S, Nishi N, Saiki I, Tokura S, Azuma I. Immunological activity of chitin and its derivatives. Vaccine. 1984;**2**(1):93-99. DOI: 10.1016/S0264-410X(98)90039-1

[98] Shigemasa Y, Saito K, Sashiwa H, Saimoto H. Enzymatic degradation of chitins and partially deacetylated chitins. International Journal of Biological Macromolecules. 1994;**16**:43-49.

[99] Tokura S, Ueno K, Miyazaki S, Nishi N. Molecular weight dependent antimicrobial activity by chitosan. Macromolecular Symposia. 1997;**120**:1-9. DOI: 10.1002/masy.19971200103

[100] Kweon DK, Song SB, Park YY. Preparation of water-soluble chitosan/heparin complex and its application as wound healing accelerator. Biomaterials. 2003;**24**:1595-1601. DOI: 10.1016/S0142-9612(02)00566-5

[101] Synowiecki J, Al-Khateeb NA. Production, properties, and some new applications of chitin and its derivatives. Critical Reviews in Food Science and Nutrition. 2003;**43**(2): 145-171. DOI: 10.1080/10408690390826473

[102] Mahae N, Chalat C, Muhamud P. Antioxidant and antimicrobial properties of chitosan-sugar complex. International Food Research Journal. 2011;**18**(4):1543-1551.

[103] Adour L, Arbia W, Amrane A, Mameri N. Combined use of waste materials – recovery of chitin from shrimp shells by lactic acid fermentation supplemented with date juice waste or glucose. Journal of Chemical Technology and Biotechnology. 2008;**83**(12): 1664-1669. DOI: 10.1002/jctb.1980

[104] Bhaskar N, Suresh PV, Sakhare PZ, Sachindra NM. Shrimp biowaste fermentation with *Pediococcus acidolactici* CFR2182: Optimization of fermentation conditions by response surface methodology and effect of optimized conditions on deproteination/demineralization and carotenoid recovery. Enzyme and Microbial Technology. 2007;**40**(5): 1427-1434. DOI: 10.1016/j.enzmictec.2006.10.019

[105] Duan S, Li L, Zhuang Z, Wu W, Hong S, Zhou J. Improved production of chitin from shrimp waste by fermentation with epiphytic lactic acid bacteria. Carbohydrate Polymers. 2012;**89**(4):1283-1288. DOI: 10.1016/j.carbpol.2012.04.051

[106] Jung WJ, Jo GH, Kuk JH, Kim KY, Park RD. Extraction of chitin from red crab shell waste by cofermentation with *Lactobacillus paracasei* subsp. *tolerans* KCTC-3074 and *Serratia marcescens* FS-3. Applied Microbiology and Biotechnology. 2006;**71**(2):234-237. DOI: 10.1007/s00253-005-0126-3

[107] Aytekin O, Elibol M. Cocultivation of *Lactococcus lactis* and *Teredinobacter turnirae* for biological chitin extraction from prawn waste. Bioprocess and Biosystems Engineering. 2010;**33**(3):393-399. DOI: 10.1007/s00449-009-0337-6

[108] Pacheco N, Garnica-González M, Ramírez-Hernández JY, Flores-Albino B, Gimeno M, Bárzana E, Shirai K. Effect of temperature on chitin and astaxanthin recoveries from shrimp waste using lactic acid bacteria. Bioresource Technology. 2009;**100**(11): 2849-2854. DOI: 10.1016/j.biortech.2009.01.019

[109] Khanafari A, Marandi R, Sanatei S. Recovery of chitin and chitosan from shrimp waste by chemical and microbial methods. Journal of Environmental Health Science & Engineering. 2008;**5**(1):1-24.

[110] Wee YJ, Kim JN, Ryu HW. Biotechnological production of lactic acid and its recent applications. Food Technology and Biotechnology. 2006;**44**(2):163-172.

[111] John RP, Nampoothiri KM, Pandey A. Fermentative production of lactic acid from biomass: an overview on process developments and future perspectives. Applied Microbiology and Biotechnology. 2007;**74**(3):524-534. DOI 10.1007/s00253-006-0779-6

[112] Litchfield JH. Microbiological production of lactic acid. Advances in Applied Microbiology. 1996;**42**:45-95.

[113] Reddy G, Altaf MD, Naveena BJ, Venkateshwar M, Kumar EV. Amylolytic bacterial lactic acid fermentation – A review. Biotechnology Advances. 2008;**26**(1):22-34. DOI: 10.1016/j.biotechadv.2007.07.004

[114] SpecialChem. [Internet]. 2014. Available from: http://www.specialchem4bio.com/news/2014/05/23/global-lactic-acid-market-to-grow-at-a-cagr-of-15-5-from-2014-20-grand-view-research#sthash.McKc8RdH.dpuf. [Accessed: January 18, 2017]

[115] Ye ZL, Lu M, Zheng Y, Li YH, Cai WM. Lactic acid production from dining-hall food waste by *Lactobacillus plantarum* using response surface methodology. Journal of Chemical Technology and Biotechnology. 2008;**83**(11):1541-1550. DOI: 10.1002/jctb.1968

[116] Probst M, Walde J, Pümpel T, Wagner AO, Insam H. A closed loop for municipal organic solid waste by lactic acid fermentation. Bioresource Technology. 2015;**175**:142-151. DOI: 10.1016/j.biortech.2014.10.034

[117] Timbuntam W, Sriroth K, Tokiwa Y. Lactic acid production from sugar-cane juice by a newly isolated *Lactobacillus* sp. Biotechnology Letters. 2006;**28**(11):811-814. DOI 10.1007/s10529-006-9003-0

[118] Dumbrepatil A, Adsul M, Chaudhari S, Khire J, Gokhale D. Utilization of molasses sugar for lactic acid production by *Lactobacillus delbrueckii* subsp. *delbrueckii* mutant Uc-3 in batch fermentation. Applied and Environmental Microbiology. 2008;**74**(1): 333-335. DOI: 10.1128/AEM.01595-07

[119] Wee YJ, Kim JN, Yun JS, Ryu HW. Utilization of sugar molasses for economical L(+)-lactic acid production by batch fermentation of *Enterococcus faecalis*. Enzyme and Microbial Technology. 2004;**35**(6):568-573. DOI: 10.1016/j.enzmictec.2004.08.008

[120] Prada-Palomo Y, Romero-Vanegas M, Díaz-Ruíz P, Molina-Velasco D, Guzmán-Luna C. Lactic acid production by *Lactobacillus* sp. from biodiesel derived raw glycerol. CT&F – Ciencia, Tecnología y Futuro. 2012;5(1):57-66

[121] Afifi MM. Enhancement of lactic acid production by utilizing liquid potato wastes. International Journal of Biological Chemistry. 2011;**5**(2):91-I02. DOI: 10.3923/ijbc.2011.91.102

[122] Mussatto SI, Dragone G, Roberto IC. Brewers' spent grain: generation, characteristics and potential applications. Journal of Cereal Science. 2006;**43**(1):1-14. DOI: 10.1016/j.jcs.2005.06.001

[123] Liguori R, Soccol CR, Vandenberghe LP, Woiciechowski AL, Ionata E, Marcolongo L, Faraco V. Selection of the strain *Lactobacillus acidophilus* ATCC 43121 and its application to brewers' spent grain conversion into lactic acid. BioMed Research International, 2015;**2015**:240231. DOI: 10.1155/2015/240231

[124] Ueno T, Ozawa Y, Ishikawa M, Nakanishi K, Kimura T. Lactic acid production using two food processing wastes, canned pineapple syrup and grape invertase, as substrate and enzyme. Biotechnology Letters. 2003;**25**(7):573-577. DOI: 10.1023/A:1022888832278

[125] Affertsholt T. Market developments and industry challenges for lactose and lactose derivatives. In: IDF Symposium "Lactose and Its Derivatives"; 14-16 May 2007, Moscow, Russia

[126] Panesar PS, Kennedy JF, Gandhi DN, Bunko K. Bioutilisation of whey for lactic acid production. Food Chemistry. 2007;**105**(1):1-14. DOI: 10.1016/j.foodchem.2007.03.035

[127] Secchi N, Giunta D, Pretti L, García MR, Roggio T, Mannazzu I, Catzeddu P. Bioconversion of ovine scotta into lactic acid with pure and mixed cultures of lactic acid bacteria. Journal of Industrial Microbiology & Biotechnology. 2012;**39**(1):175-181. DOI: 10.1007/s10295-011-1013-9

[128] Pintado J, Guyot JP, Raimbault M. Lactic acid production from mussel processing wastes with an amylolytic bacterial strain. Enzyme and Microbial Technology. 1999:**24**(8-9):590-598. DOI: 10.1016/S0141-0229(98)00168-9

[129] Kurbanoglu EB, Kurbanoglu NI. Utilization for lactic acid production with a new acid hydrolysis of ram horn waste. FEMS Microbiology Letters. 2003;**225**(1):29-34. DOI: 10.1016/S0378-1097(03)00472-5

[130] Busairi AM. Conversion of pineapple juice waste into lactic acid in batch and fed-batch fermentation systems. Reaktor. 2008;**12**(2):98-101. DOI: 10.14710/reaktor.12.2.98-101

[131] Palaniraj R, Nagarajan P. Kinetic studies in production of lactic acid from waste potato starch using *Lactobacillus casei*. International Journal of ChemTech Research. 2012;**4**(4):1601-1614.

[132] Wang L, Zhao B, Liu B, Yang C, Yu B, Li Q, Ma C, Xu P, Ma Y. Efficient production of L-lactic acid from cassava powder by *Lactobacillus rhamnosus*. Bioresource Technology. 2010;**101**(20):7895-7901. DOI: 10.1016/j.biortech.2010.05.018

[133] Chan-Blanco Y, Bonilla-Leiva AR, Velazquez AC. Using banana to generate lactic acid through batch process fermentation. Applied Microbiology and Biotechnology. 2003;**63**(2):147-152. DOI: 10.1007/s00253-003-1374-8

[134] Wang XM, Wang QH, Ren NQ, Wang XQ. Lactic acid production from kitchen waste with a newly characterized strain of *Lactobacillus plantarum*. Chemical and Biochemical Engineering Quarterly. 2005;**19**(4):383-389.

[135] Gao MT, Hirata M, Toorisaka E, Hano T. Acid-hydrolysis of fish wastes for lactic acid fermentation. Bioresource Technology. 2006;**97**(18):2414-2420. DOI: 10.1016/j.biortech.2005.10.002

[136] Abdel-Rahman MA, Tashiro Y, Sonomoto K. Lactic acid production from lignocellulose-derived sugars using lactic acid bacteria: Overview and limits. Journal of Biotechnology. 2011;**156**(4):286-301. DOI: 10.1016/j.jbiotec.2011.06.017

[137] Guo W, Jia W, Li Y, Chen S. Performances of *Lactobacillus brevis* for producing lactic acid from hydrolysate of lignocellulosics. Applied Biochemistry and Biotechnology. 2010;**161**(1-8):124-136. DOI: 10.1007/s12010-009-8857-8

[138] Martin MA, Siles JA, Chica AF, Martin A. Biomethanization of orange peel waste. Bioresource Technology. 2010;**101**(23):8993-8999. DOI: 10.1016/j.biortech.2010.06.133

[139] Liang S, Gliniewicz K, Gerritsen AT, McDonald AG. Analysis of microbial community variation during the mixed culture fermentation of agricultural peel wastes to produce lactic acid. Bioresource Technology. 2016;**208**:7-12. DOI: 10.1016/j.biortech.2016.02.054.

[140] Sreenath HK, Moldes AB, Koegel RG, Straub RJ. Lactic acid production by simultaneous saccharification and fermentation of alfalfa fiber. Journal of Bioscience and Bioengineering. 2001;**92**(6):518-523.

[141] Gullón B, Yáñez R, Alonso JL, Parajó JC. L-lactic acid production from apple pomace by sequential hydrolysis and fermentation. Bioresource Technology. 2008;**99**(2):308-319. DOI: 10.1016/j.biortech.2006.12.018

[142] Romaní A, Yáñez R, Garrote G, Alonso JL. SSF production of lactic acid from cellulosic biosludges. Bioresource Technology. 2008;**99**(10):4247-4254. DOI: 10.1016/j.biortech.2007.08.051

[143] Wee YJ, Yun JS, Park DH, Ryu HW. Biotechnological production of L(+)-lactic acid from wood hydrolyzate by batch fermentation of *Enterococcus faecalis*. Biotechnology Letters. 2004;**26**(1):71-74. DOI: 10.1023/B:BILE.0000009464.23026.e0

[144] Abe SI, Takagi M. Simultaneous saccharification and fermentation of cellulose to lactic acid. Biotechnology and Bioengineering. 1991;**37**(1):93-96. DOI: 10.1002/bit.260370113

[145] McCaskey TA., Zhou SD, Britt SN, Strickland R. Bioconversion of municipal solid waste to lactic acid by *Lactobacillus* species. Applied Biochemistry and Biotechnology. 1994;**45**(1):555-568. DOI: 10.1007/BF02941830

[146] Ghosh MK, Ghosh UK. Utilization of pine needles as bed material in solid state fermentation for production of lactic acid by *Lactobacillus* strains. BioResources. 2011;**6**(2):1556-1575.

[147] Marques S, Santos JA, Gírio FM, Roseiro JC. Lactic acid production from recycled paper sludge by simultaneous saccharification and fermentation. Biochemical Engineering Journal. 2008;**41**(3):210-216. DOI: 10.1016/j.bej.2008.04.018

[148] Laopaiboon P, Thani A, Leelavatcharamas V, Laopaiboon L. Acid hydrolysis of sugarcane bagasse for lactic acid production. Bioresource Technology. 2010;**101**(3):1036-1043. DOI: 10.1016/j.biortech.2009.08.091

[149] Nguyen CM, Kim JS, Nguyen TN, Kim SK, Choi GJ, Choi YH, Jang KS, Kim JC. Production of L-and D-lactic acid from waste *Curcuma longa* biomass through simultaneous saccharification and cofermentation. Bioresource Technology. 2013;**146**:35-43. DOI: 10.1016/j.biortech.2013.07.035

[150] Bustos G, Moldes AB, Cruz JM, Domínguez JM. Production of fermentable media from vine-trimming wastes and bioconversion into lactic acid by *Lactobacillus pentosus*. Journal of the Science of Food and Agriculture. 2004;**84**(15):2105-2112. DOI: 10.1002/jsfa.1922

[151] Yáñez R, Alonso JL, Parajó JC. D-Lactic acid production from waste cardboard. Journal of Chemical Technology and Biotechnology. 2005;**80**(1):76-84. DOI: 10.1002/jctb.1160

[152] Garde A, Jonsson G, Schmidt AS, Ahring BK. Lactic acid production from wheat straw hemicellulose hydrolysate by *Lactobacillus pentosus* and *Lactobacillus brevis*. Bioresource Technology. 2002;**81**(3):217-223. DOI: 10.1016/S0960-8524(01)00135-3

[153] Moldes AB, Alonso JL, Parajo JC. Multi-step feeding systems for lactic acid production by simultaneous saccharification and fermentation of processed wood. Bioprocess Engineering. 2000;**22**(2):175-180. DOI: 10.1007/s004490050028

[154] Rahimnejad M, Adhami A, Darvari S, Zirepour A, Oh SE. Microbial fuel cell as new technology for bioelectricity generation: A review. Alexandria Engineering Journal. 2015;**54**(3):745-756. DOI: 10.1016/j.aej.2015.03.031

[155] Kim GT, Hyun MS, Chang IS, Kim HJ, Park HS, Kim BH, Kim SD, Wimpenny JW, Weightman AJ. Dissimilatory Fe(III) reduction by an electrochemically active lactic acid bacterium phylogenetically related to *Enterococcus gallinarum* isolated from submerged soil. Journal of Applied Microbiology. 2005;**99**(4):978-987. DOI: 10.1111/j.1365-2672.2004.02514.x

[156] Freguia S, Masuda M, Tsujimura S, Kano K. *Lactococcus lactis* catalyses electricity generation at microbial fuel cell anodes via excretion of a soluble quinone. Bioelectrochemistry. 2009;**76**(1):14-18. DOI: 10.1016/j.bioelechem.2009.04.001

[157] Rosenbaum MA, Bar HY, Beg QK, Segrè D, Booth J, Cotta MA, Angenent LT. *Shewanella oneidensis* in a lactate-fed pure-culture and a glucose-fed co-culture with *Lactococcus lactis* with an electrode as electron acceptor. Bioresource Technology. 2011;**102**(3): 2623-2628. DOI: 10.1016/j.biortech.2010.10.033

[158] Arbianti R, Utami T, Hermansyah H, Novitasari D, Kristin E, Trisnawati I. Performance optimization of Microbial Fuel Cell (MFC) using *Lactobacillus bulgaricus*. Makara Journal of Technology. 2013;**17**(1):32-38. DOI: 10.7454/mst.v17i1.1925

[159] Jingna Y, Yudi S, Yi G. Property and mechanism of electricity production on lactic acid bacteria Microbial Fuel Cells. Chemical Industry and Engineering. 2014;**5**:008.

[160] Wang M, Yang Z, Xia M, Fan L, Xuejun Z, Wei S, Zou T. Performance improvement of microbial fuel cells by lactic acid bacteria and anode modification. Environmental Engineering Science. 2017;**34**(4):251-257. DOI: 10.1089/ees.2016.0110

[161] Rago L, Baeza JA, Guisasola A. Bioelectrochemical hydrogen production with cheese whey as sole substrate. Journal of Chemical Technology and Biotechnology. 2017;**92**(1):173-179. DOI: 10.1002/jctb.4987

[162] Sikora A, Błaszczyk M, Jurkowski M, Zielenkiewicz U. Lactic acid bacteria in hydrogen-producing consortia: on purpose or by coincidence? In: Kongo M, editor. Lactic Acid Bacteria – R & D for Food, Health and Livestock Purposes. Rijeka: InTech; 2013. pp. 488-514. DOI: 10.5772/50364

[163] Hung CH, Lee KS, Cheng LH, Huang YH, Lin PJ, Chang JS. Quantitative analysis of a high-rate hydrogen-producing microbial community in anaerobic agitated granular sludge bed bioreactors using glucose as substrate. Applied Microbiology and Biotechnology. 2007;**75**(3):693-701. DOI: 10.1007/s00253-007-0854-7

[164] Yang P, Zhang R, McGarvey JA, Benemann JR. Biohydrogen production from cheese processing wastewater by anaerobic fermentation using mixed microbial communities. International Journal of Hydrogen Energy. 2007;**32**(18):4761-4771. DOI: 10.1016/j.ijhydene.2007.07.038

[165] Chojnacka A, Błaszczyk MK, Szczęsny P, Nowak K, Sumińska M, Tomczyk-Żak K, Zielenkiewicz U, Sikora A. Comparative analysis of hydrogen-producing bacterial biofilms and granular sludge formed in continuous cultures of fermentative bacteria. Bioresource Technology. 2011;**102**(21):10057-10064. DOI: 10.1016/j.biortech.2011.08.063

[166] Kawaguchi H, Hashimoto K, Hirata K, Miyamoto K. H_2 production from algal biomass by a mixed culture of *Rhodobium marinum* A-501 and *Lactobacillus amylovorus*. Journal of Bioscience and Bioengineering. 2001;**91**(3):277-282. DOI: 10.1016/S1389-1723(01)80134-1

[167] Baghchehsaraee B, Nakhla G, Karamanev D, Margaritis A. Effect of extrinsic lactic acid on fermentative hydrogen production. International Journal of Hydrogen Energy. 2009;**34**(6):2573-2579. DOI: 10.1016/j.ijhydene.2009.01.010

[168] Matsumoto M, Nishimura Y. Hydrogen production by fermentation using acetic acid and lactic acid. Journal of Bioscience and Bioengineering. 2007;**103**(3):236-241. DOI: 10.1263/jbb.103.236

[169] Fedorov AS, Tsygankov AA, Rao KK, Hall DO. Hydrogen photoproduction by *Rhodobacter sphaeroides* immobilised on polyurethane foam. Biotechnology Letters. 1998;**20**(11):1007-1009. DOI: 10.1023/A:1005402904462

[170] Fascetti E, Todini O. *Rhodobacter sphaeroides* RV cultivation and hydrogen production in a one-and two-stage chemostat. Applied Microbiology and Biotechnology. 1995;**44**(3-4): 300-305. DOI: 10.1007/BF00169920

[171] Noike T, Takabatake H, Mizuno O, Ohba M. Inhibition of hydrogen fermentation of organic wastes by lactic acid bacteria. International Journal of Hydrogen Energy. 2002;**27**(11-12):1367-1371. DOI: 10.1016/S0360-3199(02)00120-9

[172] Gomes BC, Rosa PRF, Etchebehere C, Silva EL, Amâncio Varesche MB. Role of homo-and heterofermentative lactic acid bacteria on hydrogen-producing reactors operated with cheese whey wastewater. International Journal of Hydrogen Energy. 2015;**40**(28):8650-8660. DOI: 10.1016/j.ijhydene.2015.05.035

[173] Ren N, Xing D, Rittmann BE, Zhao L, Xie T, Zhao X. Microbial community structure of ethanol type fermentation in bio-hydrogen production. Environmental Microbiology. 2007;**9**(5):1112-1125. DOI: 10.1111/j.1462-2920.2006.01234.x

[174] Jo JH, Jeon CO, Lee DS, Park JM. Process stability and microbial community structure in anaerobic hydrogen-producing microflora from food waste containing kimchi. Journal of Biotechnology. 2007;**131**(3):300-308. DOI: 10.1016/j.jbiotec.2007.07.492

[175] Wang X, Zhao YC. A bench scale study of fermentative hydrogen and methane production from food waste in integrated two-stage process. International Journal of Hydrogen Energy. 2009;**34**(1):245-254. DOI: 10.1016/j.ijhydene.2008.09.100

[176] Herrmann C, Heiermann M, Idler C. Effects of ensiling, silage additives and storage period on methane formation of biogas crops. Bioresource Technology. 2011:**102**(8):5153-5161. DOI :10.1016/j.biortech.2011.01.012

[177] Xu R, Zhang BY, Yang FY. Effects of silage additives on biogas production of hybrid penisetum. Advanced Materials Research. 2015;**1070-1072**:112-120. DOI: 10.4028/www.scientific.net/AMR.1070-1072.112

[178] Lehtomäki, A. Biogas Production from Energy Crops and Crop Residues. Finland: University of Jyväskylä; 2006. 91 p.

[179] McEniry J, Allen E, Murphy JD, O'Kiely P. Grass for biogas production: The impact of silage fermentation characteristics on methane yield in two contrasting biomethane potential test systems. Renewable Energy. 2014;**63**:524-530. DOI: 10.1016/j.renene.2013.09.052

[180] Haag NL, Nägele HJ, Fritz T, Oechsner H. Effects of ensiling treatments on lactic acid production and supplementary methane formation of maize and amaranth – An advanced green biorefining approach. Bioresource Technology. 2015;**178**:217-225. DOI: 10.1016/j.biortech.2014.08.048

[181] Pakarinen O, Lehtomäki A, Rissanen S, Rintala J. Storing energy crops for methane production: Effects of solids content and biological additive. Bioresource Technology. 2008;**99**(15):7074-7082. DOI: 10.1016/j.biortech.2008.01.007

[182] Kreuger E, Nges IA, Björnsson L. Ensiling of crops for biogas production: Effects on methane yield and total solids determination. Biotechnology for Biofuels. 2011;**4**:44. DOI: 10.1186/1754-6834-4-44

[183] Menardo S, Balsari P, Tabacco E, Borreani G. Effect of conservation time and the addition of lactic acid bacteria on the biogas and methane production of corn stalk silage. BioEnergy Research. 2015;**8**(4):1810-1823. DOI: 10.1007/s12155-015-9637-7

[184] Przybył J, Kot W, Wojcieszak D, Mioduszewska N, Durczak K. Biogas yield of maize straw. Inżynieria Rolnicza. 2013;**17**(148):103-111.

[185] Zhao N, Yu M, Wang Q, Song N, Che S, Wu C, Sun X. Effect of ethanol and lactic acid prefermentation on putrefactive bacteria suppression, hydrolysis, and methanogenesis of food waste. Energy & Fuels. 2016;**30**(4):2982-2989. DOI: 10.1021/acs.energyfuels.5b02779

[186] Zhang B, Cai WM, He PJ. Influence of lactic acid on the two-phase anaerobic digestion of kitchen wastes. Journal of Environmental Sciences. 2007;**19**(2):244-249. DOI: 10.1016/S1001-0742(07)60040-0

[187] Pahlow G, Muck RE, Driehuis F, Oude Elferink SJWH, Spoelstra SF. Microbiology of ensiling. In: Buxton DR, Muck RE, Harrison JH, editors. Silage Science and Technology. Agronomy Monograph 42. Madison: American Society of Agronomy; 2003. pp. 31-93. DOI: 10.2134/agronmonogr42.c2

[188] Balat M, Balat H. Recent trends in global production and utilization of bio-ethanol fuel. Applied Energy. 2009;**86**(11):2273-2282. DOI: 10.1016/j.apenergy.2009.03.015

[189] Gold RS, Meagher MM, Hutkins R, Conway T. Ethanol tolerance and carbohydrate metabolism in lactobacilli. Journal of Industrial Microbiology. 1992;**10**(1):45-54. DOI: 10.1007/BF01583633

[190] Schell DJ, Dowe N, Ibsen KN, Riley CJ, Ruth MF, Lumpkin RE. Contaminant occurrence, identification and control in a pilot-scale corn fiber to ethanol conversion process. Bioresource Technology. 2007;**98**(15):2942-2948. DOI: 10.1016/j.biortech.2006.10.002

[191] Lucena BT, dos Santos BM, Moreira JL, Moreira JL, Nunes AC, Azevedo V, Miyoshi A, Thompson FB, de Morais MA Jr. Diversity of lactic acid bacteria of the bioethanol process. BMC Microbiology. 2010;**10**(1):298. DOI: 10.1186/1471-2180-10-298

[192] Narendranath NV, Thomas KC, Ingledew WM. Acetic acid and lactic acid inhibition of growth of *Saccharomyces cerevisiae* by different mechanisms. Journal of the American Society of Brewing Chemists. 2001;**59**(4):187-194. DOI: 10.1094/ASBCJ-59-0187

[193] Liu S, Skinner-Nemec KA, Leathers TD. *Lactobacillus buchneri* strain NRRL B-30929 converts a concentrated mixture of xylose and glucose into ethanol and other products. Journal of Industrial Microbiology & Biotechnology. 2008;**35**(2):75-81. DOI: 10.1007/s10295-007-0267-8

[194] Liu S, Nichols NN, Dien BS, Cotta MA. Metabolic engineering of a *Lactobacillus plantarum* double Ldh knockout strain for enhanced ethanol production. Journal of Industrial Microbiology and Biotechnology. 2006;**33**(1):1-7. DOI: 10.1007/s10295-005-0001-3

[195] Nair NU, Shao Z, Zhao H, Sullivan RP, McLachlan M, Johannes T, Zhao H. Biobutanol from yeast: A synergistic genome and protein engineering approach. In: The 2008 AIChE Annual Meeting, Conference Proceedings, Philadelphia, PA, USA. November 16-21, 2008.

[196] Atsumi S, Cann AF, Connor MR, Shen CR, Smith KM, Brynildsen MP, Chou KJ, Hanai T, Liao JC. Metabolic engineering of *Escherichia coli* for 1-butanol production. Metabolic Engineering. 2008;**10**(6):305-311. DOI: 10.1016/j.ymben.2007.08.003

[197] Nielsen DR, Leonard E, Yoon SH, Tseng HC, Yuan C, Prather KL. Engineering alternative butanol production platforms in heterologous bacteria. Metabolic Engineering. 2009;**11**(4-5):262-273. DOI: 10.1016/j.ymben.2009.05.003

[198] Berezina OV, Zakharova NV, Brandt A, Yarotsky SV, Schwarz WH, Zverlov VV. Reconstructing the clostridial n-butanol metabolic pathway in *Lactobacillus brevis*. Applied Microbiology and Biotechnology. 2010;**87**(2):635-646. DOI: 10.1007/s00253-010-2480-z

[199] Liu S, Bischoff KM, Qureshi N, Hughes SR, Rich JO. Functional expression of the thiolase gene thl from *Clostridium beijerinckii* P260 in *Lactococcus lactis* and *Lactobacillus buchneri*. New Biotechnology. 2010;**27**(4):283-288. DOI: 10.1016/j.nbt.2010.03.007

[200] Ghiaci P, Lameiras F, Norbeck J, Larsson C. Production of 2-butanol through meso-2,3-butanediol consumption in lactic acid bacteria. FEMS Microbiology Letters. 2014;**360**(1):70-75. DOI: 10.1111/1574-6968.12590

[201] Steinkraus KH. Introduction to indigenous fermented foods. In: Steinkraus KH, editor. Handbook of Indigenous Fermented Foods. New York: Marcel Dekker; 1995. pp. 1-6.

[202] Liu SN, Han Y, Zhou ZJ. Lactic acid bacteria in traditional fermented Chinese foods. Food Research International. 2011;**44**:643-651. DOI: 10.1016/j.foodres.2010.12.034

[203] Canadian Dairy Commission. Fermented Milk Products. 2011. http://www.milkingredients.ca/index-eng.php?id=180 [Accessed: April 6, 2017].

[204] Bylund G. Dairy Processing Handbook. Tetra Pak International S.A.; 2015. 486 p.

[205] Nip WK. (2007). Fermented products and their manufacture. In: Hui YH, editor. Handbook of Food Products Manufacturing. USA: CRC Press/Taylor & Francis; 2007. pp. 45–84. DOI: 10.1002/9780470113554.ch2

[206] Laiño JE, Juarez del Valle M, Savoy de Giori G, LeBlanc JG. Applicability of a *Lactobacillus amylovorus* strain as co-culture for natural folate bio-enrichment of fermented milk. International Journal of Food Microbiology. 2014;**191**:10-16. DOI: 10.1016/j.ijfoodmicro.2014.08.031

[207] Capozzi V, Menga V, Digesu AM, De Vita P, van Sinderen D, Cattivelli L, Fares C, Spano G. Biotechnological production of vitamin B2-enriched bread and pasta. Journal of Agricultural and Food Chemistry. 2011;**59**(14):8013-8020. DOI: 10.1021/jf201519h

[208] Taranto MP, Vera JL, Hugenholtz J, De Valdez GF, Sesma F. *Lactobacillus reuteri* CRL1098 produces cobalamin. Journal of Bacteriology. 2003;**185**(18):5643-5647. DOI: 10.1128/JB.185.18.5643-5647.2003

[209] Morishita T, Tamura N, Makino T, Kudo S. Production of menaquinones by lactic acid bacteria. Journal of Dairy Science. 1999;**82**(9):1897-1903. DOI: 10.3168/jds.S0022-0302(99)75424-X

[210] Davies EA, Bevis HE, Delves-Broughton J. The use of the bacteriocin, nisin, as a preservative in ricotta-type cheeses to control the food-borne pathogen *Listeria monocytogenes*. Letters in Applied Microbiology. 1997;**24**(5):343-346. DOI: 10.1046/j.1472-765X.1997.00145.x

[211] Pei J, Yue T, Yuan Y, Dai L. Activity of paracin C from lactic acid bacteria against *Alicyclobacillus* in apple juice: Application of a novelty bacteriocin. Journal of Food Safety. 2017;00:e12350. DOI: 10.1111/jfs.12350

[212] Miao J, Guo H, Ou Y, Liu G, Fang X, Liao Z, Ke C, Chen Y, Zhao L, Cao Y. Purification and characterization of bacteriocin F1, a novel bacteriocin produced by *Lactobacillus paracasei* subsp. tolerans FX-6 from Tibetan kefir, a traditional fermented milk from Tibet, China. Food Control. 2014;**42**:48-53. DOI: 10.1016/j.foodcont.2014.01.041

[213] Nielsen JW, Dickson JS, Crouse JD. Use of a bacteriocin produced by *Pediococcus acidilactici* to inhibit *Listeria monocytogenes* associated with fresh meat. Applied and Environmental Microbiology. 1990;**56**(7):2142-2145

[214] Sanni AI, Onilude AA, Ogunbanwo ST, Smith SI. Antagonistic activity of bacteriocin produced by *Lactobacillus* species from ogi, an indigenous fermented food. Journal of Basic Microbiology. 1999;**39**(3):189-195. DOI: 10.1002/(SICI)1521-4028(199906)39:33.3.CO;2-I

[215] Sonsa-Ard N, Rodtong S, Chikindas ML, Yongsawatdigul J. Characterization of bacteriocin produced by *Enterococcus faecium* CN-25 isolated from traditionally Thai fermented fish roe. Food Control. 2015;**54**:308-316. DOI: 10.1016/j.foodcont.2015.02.010

[216] Rodríguez-Alcalá LM, Braga T, Malcata FX, Gomes A, Fontecha J. Quantitative and qualitative determination of CLA produced by *Bifidobacterium* and lactic acid bacteria by combining spectrophotometric and Ag+-HPLC techniques. Food Chemistry. 2011;**125**(4):1373-1378. DOI: 10.1016/j.foodchem.2010.10.008

[217] Herzallah S. Enrichment of conjugated linoleic acid (CLA) in hen eggs and broiler chickens meat by lactic acid bacteria. British Poultry Science. 2013;**54**(6):747-752. DOI: 10.1080/00071668.2013.836734

[218] Liu P, Shen SR, Ruan H, Zhou Q, Ma LL, He GQ. Production of conjugated linoleic acids by *Lactobacillus plantarum* strains isolated from naturally fermented Chinese pickles. Journal of Zhejiang University. Science. B. 2011;**12**(11):923-930. DOI: 10.1631/jzus.B1100072

[219] Duboc P, Mollet B. Applications of exopolysaccharides in the dairy industry. International Dairy Journal. 2001;**11**(9):759-768. DOI: 10.1016/S0958-6946(01)00119-4

[220] Han X, Yang Z, Jing X, Yu P, Zhang Y, Yi H, Zhang L. Improvement of the texture of yogurt by use of exopolysaccharide producing lactic acid bacteria. BioMed Research International. 2016;**2016**:1-6. DOI: 10.1155/2016/7945675

[221] Low D, Ahlgren JA, Horne D, McMahon DJ, Oberg CJ, Broadbent JR. Role of *Streptococcus thermophilus* MR-1C capsular exopolysaccharide in cheese moisture retention. Applied and Environmental Microbiology. 1998;**64**(6):2147-2151.

[222] Macura D, Townsley PM. Scandinavian ropy milk: Identification and characterization of endogenous ropy lactic streptococci and their extracellular excretion. Journal of Dairy Science. 1984;**67**:735-744.

[223] Adapa S, Schmidt KA. Physical properties of low-fat sour cream containing exopolysaccharide producing lactic acid. Journal of Food Science. 1998;**63**:901-903. DOI: 10.1111/j.1365-2621.1998.tb17922.x

[224] Hong SH, Marshall RT. Natural exopolysaccharides enhance survival of lactic acid bacteria in frozen dairy desserts. Journal of dairy science. 2001;**84**(6):1367-1374. DOI: 10.3168/jds.S0022-0302(01)70167-1

[225] Di Cagno R, De Angelis M, Limitone A, Minervini F, Carnevali P, Corsetti A, Gaenzle M, Ciati R, Gobbetti M. Glucan and fructan production by sourdough *Weissella cibaria* and *Lactobacillus plantarum*. Journal of Agricultural and Food Chemistry. 2006;**54**(26):9873-9881. DOI: 10.1021/jf061393+

[226] Rodríguez C, Medici M, Rodríguez AV, Mozzi F, Font de Valdez G. Prevention of chronic gastritis by fermented milks made with exopolysaccharide-producing *Streptococcus thermophilus* strains. Journal of Dairy Science. 2009;**92**(6):2423-2434. DOI: 10.3168/jds.2008-1724

[227] Tok E, Aslim B. Cholesterol removal by some lactic acid bacteria that can be used as probiotic. Microbiology and Immunology. 2010;**54**(5):257-264. DOI: 10.1111/j.1348-0421.2010.00219.x

[228] Beckman KB, Ames BN. The free radical theory of aging matures. Physiological Reviews. 1998;**78**(2):547-581.

[229] Virtanen T, Pihlanto A, Akkanen S, Korhonen H. Development of antioxidant activity in milk whey during fermentation with lactic acid bacteria. Journal of Applied Microbiology. 2007;**102**:106-115. DOI: 10.1111/j.1365-2672.2006.03072.x

[230] Pyo YH, Lee TC, Lee YC. Effect of lactic acid fermentation on enrichment of antioxidant properties and bioactive isoflavones in soybean. Journal of Food Science. 2005;**70**: S215–S220. DOI: 10.1111/j.1365-2621.2005.tb07160.x

[231] Ravyts F, De Vuyst L. Prevalence and impact of single-strain starter cultures of lactic acid bacteria on metabolite formation in sourdough. Food microbiology. 2011;**28**(6):1129-1139. DOI: 10.1016/j.fm.2011.03.004

[232] Soetaert W, Vanhooren PT, Vandamme EJ. Production of mannitol by fermentation. Carbohydrate Biotechnology Protocols. 1999;**10**:261-275. DOI: 10.1007/978-1-59259-261-6_21

[233] Saha BC, Racine FM. Biotechnological production of mannitol and its applications. Applied Microbiology and Biotechnology. 2011;**89**(4):879-891. DOI: 10.1007/s00253-010-2979-3

[234] Monedero V, Pérez-Martínez G, Yebra MJ. Perspectives of engineering lactic acid bacteria for biotechnological polyol production. Applied Microbiology and Biotechnology. 2010;**86**(4):1003-1015. DOI: 10.1007/s00253-010-2494-6

Chapter 6

Food Additives and Processing Aids used in Breadmaking

Luis Carlos Gioia, José Ricardo Ganancio and
Caroline Joy Steel

Additional information is available at the end of the chapter

http://dx.doi.org/10.5772/intechopen.70087

Abstract

The main classes of additives used in breadmaking are: (i) oxidants/reductants; (ii) emulsifiers; (iii) hydrocolloids; and (iv) preservatives. The main processing aids used are enzymes. Historically, market trends have developed from the use of ingredients in greater quantities - to obtain specific effects in bread (such as fat for crumb softness) - to the use of additives at much lower levels (max. 1%) and, more recently, to enzymes which are used in parts per million (ppm). According to many regulations, enzymes do not need to be declared on the label of the final product, attending the "clean label" trend. We will describe the food additives used under each class, individually describing their mode of action and effects on dough rheology, during the breadmaking process, and on product quality. We will also describe the main enzymes currently used, dividing them according to the substrate they act on (gluten, starch, lipids, non-starch polysaccharides or NSPS), individually describing their mode of action and effects on dough rheology, during the breadmaking process, and on product quality. Legal aspects will also be addressed. We will conclude with future trends in the use of additives and processing aids in breadmaking.

Keywords: bread, oxidants, reductants, emulsifiers, hydrocolloids, preservatives, enzymes

1. Additives in breadmaking

The main classes of additives used in breadmaking are: (i) oxidants/reductants; (ii) emulsifiers; (iii) hydrocolloids; and (iv) preservatives. Maximum dosages permitted may vary according to the application and from country to country; so local legislation must always be consulted. Usually, the Joint FAO/WHO Expert Committee on Food Additives (JECFA) of the Codex Alimentarius, the Food and Drug Administration (FDA) and the European Food Safety Authority (EFSA) are taken as guides. The International Numbering System, created in the European Union, assigns E-numbers to all approved food additives, and these are used in many countries to facilitate identification.

1.1. Oxidants and reductants

Oxidants and reductants are normally included to assist with gluten network development [1]. Oxidants improve stability and elasticity of the dough, which becomes stronger, increasing oven rise, and making crumb grain finer. They act on the gluten proteins of flour, i.e. oxidizable thiol (–SH) groups, creating additional disulfide bonds (S-S) [2]. Oxidative enzymes such as glucose-oxidase and hexose-oxidase are now used to replace or support the action of traditional redox materials [3]. Reductants have the opposite effect, but may help to optimize gluten network formation.

1.1.1. Azodicarbonamide (ADA) (E927)

Azodicarbonamide (ADA) is a fast-acting oxidizing agent. Its action is to oxidize free thiol groups (– SH) in flour proteins and to strengthen the dough. This action is particularly effective in modifying the dough properties of poor-quality flours, for instance by improving the processing behavior and gas retention properties. ADA used at the correct level increases bread volume and improves crumb properties, but overdosing depresses loaf volume [4].

Azodicarbonamide is a maturing agent used in flour premixes, providing immediate oxidation when water is added. It is consumed in the mixer, in the early stages of the baking process. Azodicarbonamide is added at dosages of 10–40 ppm (flour basis) [4].

The use of ADA is banned in EU countries, but is still used in others. The key reason for the ban is the presence of a reaction product, semicarbazide, which is present in bread crumb and crust, posing a health risk. The use of oxidizing agents depends on legislation, flour quality and production process. In European countries, only ascorbic acid is permitted [4].

1.1.2. Ascorbic acid (E300)

Ascorbic acid is commonly used as an improver in the baking industry. In some countries, it is the only oxidation improver allowed. It has an intermediate speed of reaction and its effect is greatly noticed in the proofing chamber. Its key mechanism of action is the sulfhydryl/disulfide reaction, which plays an important role in the rheological properties of bakery systems [3].

Ascorbic acid itself is a reducing agent. However, in the presence of oxygen and an enzyme, ascorbic acid-oxidase, which is naturally found in wheat flour, it is converted to its dehydro form, that participates in oxidation reactions, stabilizing the gluten network [4]. Its effect on gluten and dough is to reduce extensibility and increase elasticity, giving better volume, shape, and finer and more uniform texture to the finished breads [5]. It is applied in pan bread from 50 to 200 ppm (flour basis) levels.

Some plants and fruits have high levels of ascorbic acid and this presents an opportunity to use them to provide the ascorbic acid requirement in bakery products. This has an advantage in that the chemically synthesized version has an E-number and must be declared on the label as ascorbic acid, vitamin C or E300, while plant or fruit products are declared as ingredients [4].

1.1.3. L-Cysteine (E920)

L-Cysteine is a reductant or reducing agent, with an inverse effect to oxidants. It is an amino acid that contains a free — SH group in its molecule, which breaks disulfide bonds between gluten-forming proteins, reducing the number of cross-links. The resulting dough is softer, lower in elasticity and greater in extensibility. L-Cysteine used alone would not be beneficial to a dough system, as it would result in bread with low volume and coarse crumb structure [4].

The advantages of using L-cysteine are improved machinability, shorter mixing time and reduced proofing time [4], a process called activated dough development (ADD). In ADD, reducing agents convert high molecular weight glutenins into smaller molecules during mixing. Extra oxidizing agents added to the dough form larger molecules again during proofing, re-establishing desired dough characteristics for breakmaking. L-Cysteine opens the disulfide bonds during mixing (less energy) while ascorbic acid closes the remaining bonds. The added oxidant must not be strong, for otherwise L-cysteine will be oxidized to cystine (dough strengthener) [2].

As L-cysteine relaxes the gluten structure during the mixing process and enhances dough development, when the dough temperature is an issue, L-cysteine may be used to reduce the work input requirement thus assisting to control the final dough temperature [5]. Its application dosage varies from 50 to 300 ppm (flour basis).

'Natural' alternatives to synthetic L-cysteine are available, which are based on inactivated yeast. In this case, the reducing effect is based on a mixture of glutathione and proteolytic enzymes released from the disrupted yeast cells [5].

1.2. Emulsifiers

Emulsifiers are common additives used in breadmaking and can be classified according to two main functions: (i) crumb softeners; and (ii) dough conditioners or gluten strengtheners. Mono- and diglycerides are the main examples of the first group, while diacetyl tartaric acid (DATA) esters of mono- and diglycerides (DATEM) and polysorbate are two prominent examples of the second. Lactylates can be classified as having both functions.

Emulsifiers are often evaluated according to their physicochemical properties. The hydrophilic/lipophilic balance concept (HLB) is the most widely used concept, although not very common in the bakery industry [6].

1.2.1. Mono- and diglycerides (E471)

Mono- and diglycerides and their derivatives account for about 70% of the production of food emulsifiers in the world. Overall, bakery is by far the field of greatest application.

Approximately, 60% of all monoglycerides are used in bakery – 40% in bread and 20% in sponge cakes and cakes [6].

Mono- and diglycerides are generally manufactured by esterification (glycerolysis) of triglycerides with glycerol, yielding a mixture of mono, di and triglycerides. The hardness of a monoglyceride is mainly determined by the hardness of the edible fat from which the monoglyceride has been produced [6]. As the monoglycerides are the functional part, molecular distillation can be carried out to increase their concentration.

The content of monoglycerides in commercially distilled monoglycerides is usually 90–95% [6]. Two crystalline forms are generally present: alpha and beta. The alpha form is the most functional type of monoglycerides in bakery products. The monoglycerides marketed for bakery applications include plastic, hydrated, powdered and distilled monoglycerides [7].

Monoglycerides possess a lipophilic character and are therefore assigned with a low HLB number (3–6). They dissolve in oil and in stabilized water-in-oil (w/o) emulsions to form reversed micelles in oil. Any functionality of monoglycerides and other emulsifiers in bakery depends on the dispersibility properties of the emulsifiers during mixing of the dough. The factors that influence dispersibility properties during dough mixing are a balance between particle size and hardness or melting point of the monoglyceride [6].

Distilled monoglycerides are considered anti-staling agents in breads, as they soften the crumb of the product after baking and retain this softness during the beginning of shelf-life. They act by binding to the amylose fraction of wheat starch at the high temperatures typical of baking. In doing so, they slow down retrogradation of the starch during cooling and subsequent storage [5].

Distilled monoglycerides have the greatest effect on softness compared to other types of emulsifiers, and less effect on loaf volume. The result is a fine crumb with considerable elasticity. The optimal dosage is 0.2% (flour basis) [2].

1.2.2. Diacetyl tartaric acid esters of mono- and diglycerides (DATEM) (E472)

DATEM include glycerol derivatives esterified with edible fatty acids and mono- and diacetyl tartaric acid [8], generally permitted for the use in foodstuffs and as dough conditioners for all baked products, particularly yeast-leavened products, such as white bread. Their HLB value is 8–10. The optimal dosage is between 0.25 and 0.50% (flour basis) [2].

DATEM comes as a sticky viscous liquid, or with a consistency like fats, or yellow waxes, or in flakes or powder form. DATEM is more hydrophilic compared to the mono- and diglycerides, and its starting materials [8].

When the flour used for breadmaking contains an inadequate amount, or less than ideal quality, of protein, the inclusion of DATEM assists in dough performance during manufacturing (tolerance toward raw material quality, mechanical resistance, sticking to manufacturing equipment, mixing and fermentation tolerance) and provides dough with reasonable oven spring [5].

Ionic emulsifiers, such as DATEM, offer a huge ability toward the formation of hydrogen bridges with amidic groups of the gluten proteins [8]. Diacetyl tartaric acid (DATA) esters bind rapidly to the hydrated gluten proteins and, as a result, the gluten network formed becomes stronger, more extensible and more resilient, producing a uniform and stable gas cell structure [5].

DATA esters enhance gas retention when incorporated into most yeast-raised wheat flour-based doughs. They have a strong improving effect on loaf volume and dough stability, which generates a more symmetrical appearance for the baked bread. Internally, breads have a finer gas cell structure with thinner cell walls, resulting in whiter crumbs, and a finer, more even texture, that is softer and more resilient [5].

For whole meal and grain breads, the major difficulty is the disruption of the gas cell network by larger particles, such as bran and seeds. This can be solved by adding extra wheat gluten, by using DATEM (or DATA esters), or by using a combination of both [5].

1.2.3. Lactylates: calcium stearoyl-lactylate (CSL) (E482) and sodium stearoyl-lactylate (SSL) (E481)

Lactylate esters are synthesized from food-grade fatty acids and lactic acid. For lactylates as emulsifiers, the fatty acid represents the non-polar portion and the ionic lactic acid polymer represents the polar portion [9].

Calcium stearoyl-lactylate (CSL) and sodium stearoyl-lactylate (SSL) are typical dough conditioners with HLB values of 8–10 and 10–12, respectively. Both are commonly used in the manufacturing of white bread and are employed as dough strengtheners. Also, they act as anti-staling agents, aeration aids and starch/protein complexing agents. Their optimal dosage is 0.25–0.50% (flour basis) [2].

Because of their high degree of hydrophilicity, lactylate salts hydrate readily in water at room temperature. The sodium salts hydrate more rapidly than the calcium salts, giving SSL and CSL different functionalities in short baking processes [9].

The strengthening effect of lactylates relates to their ability to aggregate proteins, which helps in the formation of the gluten matrix. It is believed that they interact with proteins through: (i) hydrophobic bonds between the non-polar regions of proteins and the stearic acid moiety of lactylates; and (ii) ionic interactions between the charged amino acid residues of proteins and the carboxylic portion of lactylates. In the case of bread dough, these effects result in increased dough viscosity, better gas retention and, ultimately, greater bread volume [9].

The effects of lactylates on dough handling properties and proofed dough volume are also related to protein complexing. As proofed dough is heated in the early baking phase, the lactylates are transferred from the protein to the starch. The coating on the starch significantly delays starch gelatinization, which keeps the viscosity low and allows additional expansion of the dough in the oven. As the resultant dough is softer than the unemulsified dough, it allows more abusive mechanical working without causing irreversible damage to the protein structure. Both CSL and SSL provide very good yeast-raised dough strengthening effects [9].

SSL enhances gas retention in the dough, but is less efficient than other dough strengthening emulsifiers, such as DATEM. It also has effects on crumb softening, extending shelf-life, through binding to amylose, showing similar action to distilled monoglycerides. However, bakers tend to prefer DATEM as a dough conditioner for maximum gas retention, and add distilled monoglycerides at the desired level when extra softness is needed [5].

SSL may be replaced by CSL at similar levels, with similar effects in breadmaking. The need to reduce sodium in bakery products, for health reasons, has led to an increased interest in CSL as an SSL replacer [5].

1.2.4. Polysorbates (E491–E496)

Polysorbates are sorbitol derivatives and they form part of a group of emulsifiers known as sorbitan esters, which can be further modified to polysorbates [10].

The polysorbate family of products is among the most hydrophilic or water soluble emulsifiers allowed in foods, due to the long polyoxyethylene chain, so the addition of small amounts of polysorbate emulsifiers to water results initially in a dramatic decrease in interfacial tension [10].

The unique qualities of each polysorbate are attributed to the different fatty acids used in each product. The ethylene oxide chain length is controlled at an average of 20 moles and it does not change between products. The short-chain fatty acid polysorbate 20 has the highest HLB at 16.7, followed by the others with longer-chains, such as polysorbates 40, 60, 65, 80 and 85 [10].

Sorbitan esters and polysorbates are emulsifiers regulated by governing bodies. For instance, in North America, the market where they are most popular, the specific applications for these compounds in foods are defined and the use level is controlled. Most polysorbates are used in bakery goods. In most bakery applications, polysorbates are used below 0.3% (flour basis) [10].

Polysorbates are added as dough strengtheners to improve baking performance. They stabilize the dough during late proofing and early stages of baking, when there are great stresses on the inflating cells. Their use results in loaves with greater volume and a fine and uniform crumb structure [10].

Regardless of its good effects in breadmaking, and the fact that the polymerized forms of ethylene oxide used in polysorbates have been shown to be safe, the unreacted free-ethylene oxide has been classified as "carcinogenic to humans (Category 1)" by the International Agency for Research on Cancer, and thus, the European Commission Scientific Committee on Food is concerned with these impurities. So, even if the potential risk of impurities in polysorbates is low, a responsible food manufacturer should be aware of these concerns. Food producers would be prudent to source their polysorbates from a reputable supplier [10].

1.3. Hydrocolloids

Hydrocolloids are widely used in the food industry, because they modify the rheology and texture of aqueous systems. These additives play a very important role in foods, as they act as stabilizers, thickeners and gelling agents, affecting the stabilization of emulsions, suspensions, and foams, and modifying starch gelatinization [2].

During baking, starch gelatinization and protein coagulation take place and the aerated structure obtained during leavening is fixed, forming the bread crumb. It has been stated that granule swelling can be reduced by the presence of hydrocolloids (particularly at high concentrations), which can interact with the molecules leached out from starch granules, leading to a stiffening effect. Thus, due to these interactions, crumb structure and texture are positively influenced by the presence of gums [11].

In the baking industry, hydrocolloids are very important as breadmaking improvers, because they enhance dough-handling properties, improve the quality of fresh bread, and extend the shelf-life of stored bread. They must be used in small quantities (<1% flour basis) and are expected to increase water retention and loaf volume, while decreasing firmness and starch retrogradation [2].

Polysaccharides such as carboxymethyl cellulose, guar gum and xanthan gum are employed as stabilizers in bakery products in particular.

1.3.1. *Xanthan gum (E415)*

Xanthan gum is an anionic polysaccharide employed to modify rheological properties of food products [1]. It is produced industrially from carbon sources through fermentation by the Gram-negative bacterium *Xanthomonas campestris* [12]. Structure-wise, it is a polymer with a D-glucose backbone. Trisaccharide side-chains formed by glucuronic acid sandwiched between two mannose units are linked to every second glucose of the main polymer chain. The carboxyl groups in xanthan gum may ionize creating negative charges, increasing the viscosity of the solution in water [1].

Xanthan gum easily disperses in cold and hot water, quickly producing viscous solutions. These solutions are stable to acid, salt, and high temperature processing conditions, and show good efficiency at low concentrations, around 0.1% (flour basis). Also, products that contain this gum have fluidity, good mouthfeel, and adhesion. These advantages make xanthan gum a suitable thickener, stabilizer, and suspending agent in many foods [12]. In bakery products, it improves wheat dough stability during proofing. Also, it has the ability to increase dough stability during freeze-thaw cycles in frozen dough [2].

1.3.2. *Guar gum (E412)*

Guar gum is made of the powdered endosperm of the seeds of *Cyamopsis tetragonolobus*, a leguminous crop. The endosperm contains a complex polysaccharide, a galactomannan, which is a polymer of D-galactose and D-mannose. This hydroxyl group-rich polymer forms hydrogen bonds with water, imparting significant viscosity and thickening to the solution. Due to its thickening, emulsifying, binding and gelling properties, quick solubility in cold water, wide pH stability, film forming ability and biodegradability, guar gum finds applications in a large number of industries, including the bakery industry. At the level of 0.5% (flour basis) in bread, it improves both softness and loaf volume. It is also used for increasing dough yield in baked goods [13].

1.3.3. Carboxymethylcellulose (CMC) (E466)

Carboxymethylcellulose (CMC) is a cellulose derivative, and it is also called cellulose gum. It finds applications in the food industry as a food stabilizer and thickener. It contains carboxymethyl groups (–CH_2COOH) attached to –OH groups within the glucopyranose monomers forming a carboxymethyl gum backbone. This anionic polysaccharide is often used as a food additive in its sodium salt form (sodium carboxymethylcellulose). In sodium carboxymethylcellulose, some of the carboxyl groups have been replaced by sodium carboxylate groups. The degree of substitution by sodium ions, chain length of the cellulose backbone and clustering of the carboxymethyl substituents determine CMC functionality [1].

CMC has a combined effect with enzymes and emulsifiers on textural properties of both dough and fresh bread. For example, CMC contributes to yielding high volume and retarding staling. Both CMC and guar gum have proven to be beneficial in the formulation of gluten-free breads [2].

1.4. Preservatives

Preservatives are intended to inhibit the growth of molds and thermophilic bacteria. The preservatives permitted for use in bread are commonly limited by legislation [5]. Propionic, sorbic and benzoic acids (E280, E202 and E210, respectively) are among the most commonly used food preservatives. Propionic acid inhibits molds and *Bacillus* spores, but not yeasts to the same extent, and has, therefore, been the traditional choice for bread preservation [14].

Preservatives are often added in their salt form, which is more soluble in aqueous solutions. Their effectiveness depends on the pH of the system to which they are added, as the dissociated acid alters the antimicrobial effect. The p*K*a values (pH at which dissociation occurs) of propionic acid and sorbic acid are 4.88 and 4.76, respectively. Maximum pH for their activity is around 6.0–6.5 and 5.0–5.5 for sorbate and propionate, respectively. At pH 6, only 7% of the propionic acid will be undissociated, compared to 71% at pH 4.5 [14].

1.4.1. Propionates

The sodium, potassium and calcium salts of propionic acid are used as bread preservatives in many countries. These preservatives have two functions: (i) to retard the rate of mold development; and (ii) to prevent the bacterial spoilage of bread known as "rope" caused by certain *Bacillus* spp., notably *B. subtilis* and *B. licheniformis*. Calcium propionate (E282) is more widely used than propionic acid, because it is easier to handle the solid salt than the corrosive liquid acid [15]. Its regular dosage is around 0.3% (flour basis).

Although effective at retarding molds and preventing "rope" spoilage, there are some practical disadvantages associated with the use of calcium propionate, among which is the effect on loaf volume. A decrease in loaf volume is caused by the combination of reduced yeast activity and altered dough rheology [15].

Regarding propionic acid, high levels of dietary intake have been associated with propionic acidemia in children. Complications of this disease can include learning disabilities, seizures, arrhythmia, gastrointestinal symptoms, recurrent infections and many others [16].

http://dx.doi.org/10.5772/intechopen.70087

1.4.2. Sorbates

Sorbates are more effective at inhibiting mold growth than propionates by weight [16]. However, sorbic acid and its salts are of less value in bread and yeast-raised goods, because of their detrimental effects on dough and bread characteristics. They can produce sticky doughs which are difficult to handle; and the baked products may have reduced volume and an irregular cell structure. The use of encapsulated sorbic acid is an alternative to overcome these negative effects. Also, sorbic acid or its salts may be sprayed on the surface of breads after baking [14]. In the dough, its dosage is around 0.1% (flour basis).

1.4.3. Acetates

Acetic acid in the form of vinegar has been used by bakers for many years to prevent the bacterial spoilage of bread known as "rope" and to increase mold-free shelf-life. It gives products a more "natural" appeal and is effective against "rope" at concentrations equivalent to 0.1–0.2% of acetic acid (flour basis). However, at such concentrations, its effect against molds is limited. Significantly higher concentrations lead to an unacceptable odor of vinegar in the bread [15].

1.4.4. Fermentates

An increasing number of natural preservatives are being marketed as "clean label" or "label friendly" shelf-life extension solutions for the bakery industry. Among these are fermentates, which are food ingredients produced by the fermentation of a variety of raw materials by food grade microorganisms. Such microorganisms include lactic acid bacteria or propionic acid bacteria that produce weak organic acids with a preservative effect. However, weak organic acid preservatives have actually been reported to have no effect on the shelf-life of bakery products with pH values close to 7 [16].

Preservatives inhibit microbial spoilage, but do not destroy microorganisms. Therefore, it is important to process baked goods following good manufacturing practices (GMP), including the use of good quality raw-materials and appropriate hygiene systems that are correctly monitored [5].

2. Enzymes in breadmaking

Enzymes, also called biocatalysts, are proteins with special properties. They are able to catalyze chemical reactions at low energy requirements without being consumed by these reactions; and the resultant effects modify the structure and/or the physicochemical properties of the environment. Each kind of enzyme has its own specific substrate on which it acts, which provides excellent process control for the use in breadmaking. As the enzymes used are not active in the final products, once they are denatured in the oven, they are classified as "processing aids", and do not need to be included in the list of ingredients in product labels, according to the legislation requirements in many countries. The Enzyme Commission (EC) number for each enzyme mentioned is shown in this chapter. This is an international numerical classification for enzymes, where classifying criteria are the chemical reactions each enzyme catalyzes [17]. For a logical comprehension, we have classified food enzymes used in baking by the substrate each one acts on, as follows.

2.1. Substrate: polysaccharides

The main polysaccharide present in wheat flour is starch, which is present in the form of granules composed of two fractions. One fraction is amylose (25–28%), the linear fraction, composed by glucose molecules linked by α-1,4 bonds; and the other fraction is amylopectin (72–75%) which is a branched fraction. Amylopectin is also a glucose polymer formed by α-1,4 bonds and branches are linked to the linear backbone by α-1,6 bonds. In the milling process, some starch granules become damaged and it is necessary to have between 7 and 11% of this damaged starch in wheat flour, once it is the substrate for α-amylase action [18–20].

2.1.1. Fungal α-amylase (EC 3.2.1.1)

This kind of endoamylase randomly hydrolyzes α-1,4 bonds of damaged starch granules from wheat flour, generating low molecular weight dextrins and oligosaccharides (maltose, maltotriose, etc.). Each generated dextrin has its own non-reducing end. Subsequently, the endogenous wheat flour β-amylase hydrolyzes generated dextrins to maltoses [19], which will be hydrolyzed to glucose by maltase enzyme produced by the yeast [18, 20].

The maximum activity pH range of fungal α-amylase varies from 5 to 6, and fits with the pH of most bread doughs [20]. Fungal α-amylases are mostly denatured by heat before starch gelatinization temperature range is reached. This fact explains why it is necessary to have damaged starch to be hydrolyzed by this enzyme: it is a more easily degradable substrate than native starch granules. There is a smaller risk of over-action of fungal α-amylase due to its lower thermostability [18].

The combined use of fungal α-amylase with endogenous β-amylase produces higher levels of maltose, stimulating yeast fermentation. Consequently, higher gas production enhancing bread volume occurs [20].

Endogenous α-amylase is present in ungerminated wheat, but its activity varies and can be indirectly measured by the Falling Number (FN). Its activity is low in ungerminated wheat, providing high FN results. On the contrary, in germinated wheat, its activity is high, causing low FN results, and this situation can be a disaster for baking. So, it is necessary to standardize flour with fungal α-amylase to guarantee the same good results in baking in terms of bread volume, crust, color and general loaf quality [18].

α-Amylase also contributes to a better crumb texture. Once it degrades damaged starch, the dough consistency decreases and machinability is enhanced [18, 20].

Another important contribution of fungal α-amylase for baking is that reducing sugars generated during mixing and fermentation will participate in the Maillard reaction (combination of low molecular weight reducing sugars with proteins under high temperature). Maillard reaction is responsible for the non-enzymatic browning of bread crust and generation of bread characteristics including aroma and flavor [18, 20].

Amylases also permit oven spring to occur for a prolonged period. The bread volume is increased once they avoid quick viscosity rising during starch gelatinization [18].

2.1.2. *β-Amylase (EC 3.2.1.2)*

This endogenous enzyme is present in mature ungerminated wheat, and hydrolyzes only damaged starch granules [18]. In breadmaking, this exo-amylase acts sequentially from the non-reducing ends of starch fractions (amylose and amylopectin) or dextrins, hydrolyzes α-1,4 bonds and releases maltoses and β-limit dextrins. The generated maltoses will be substrate for yeast fermentation after maltase action, enhancing the gassing power of the dough [19]. β-Amylase action ceases one glucose molecule before an α-1,6 bond of amylopectin. The α-1,6 bond is the branching point of amylopectin [20]. This effect also contributes to reduce bread firmness [18]. The maltoses generated that are not consumed by the yeast contribute to crust color [19].

2.1.3. *Bacterial amylase*

This enzyme hydrolyzes starch more aggressively than fungal α-amylase. This effect is due to its efficiency to act on amorphous regions of starch granules, generating excessive dextrinization, with excessive decrease in dough viscosity, producing an open texture crumb [20].

Bacterial amylase provides a softer crumb, despite greater recrystallized starch content in comparison with a control. However, stickiness and gumminess were verified in crumb treated with this enzyme. Such effect occurs by greater thermostability of bacterial amylase, which keeps its capacity to hydrolyze gelatinized starch inside the oven, when fungal α-amylase is already denatured, and its action may continue during storage [18, 20].

It was proven that bacterial amylase was efficient to extend bread shelf-life. However, small overdosing provokes great and undesirable texture modification [20].

2.1.4. *Bacterial maltogenic α-amylase (EC 3.2.1.133)*

Bacterial maltogenic α-amylase is obtained from genetically modified *Bacillus stearothermophilus*. This enzyme hydrolyzes α-1,4 linkages of easily accessible outer gelatinized starch molecules, in both amylose and amylopectin fractions, producing α-maltose and other malto-oligosaccharides [21], decreasing bread staling speed. The hydrolyzed amylopectin branches project themselves to the intergranular spaces hampering their reorganization, avoiding crystallization and/or amylose-amylopectin interactions, providing a weaker and less firm starch structure, yielding softer bread [18].

This exo-enzyme is unable to hydrolyze α-1,6 linkages, so it stops its action one glucose molecule before starch branching. Also, there are some evidences of endo-activity, shown by amylose and β-limit dextrin hydrolysis. The lower molecular weight branched oligosaccharides resulting from maltogenic α-amylase action on amylopectin, maltotriose and/or maltotetralose, act as anti-firming agents in baked goods [18, 22].

According to Gerrard et al. [23], the use of maltogenic α-amylase did not affect rheological properties of bread dough due to its low activity at mixing temperatures (lower than 35°C). Its higher activity is observed at starch gelatinization temperatures during the baking stage, which is enough for the hydrolysis of glycosidic bonds in gelatinized starch by this enzyme. The inactivation of this enzyme by high temperatures occurs during baking time, and starch hydrolysis produces a limited amount of soluble dextrins.

The produced maltodextrins inhibit starch-starch and starch-protein interactions causing a delay in amylopectin reassociation and retrogradation, resulting in a slower crumb firming process. This effect is known as anti-staling [18].

2.1.5. Amyloglucosidase or glucoamylase (EC 3.2.1.3)

This exo-amylase directly releases α-glucose molecules from native or damaged starch granules, increasing the production rate of fermentable sugars in the dough, enhancing yeast fermentation rate [18]. The level of added sugars can be reduced by using amyloglucosidase, and crust color can be improved, as enzyme activity remains after yeast inactivation. As glucose continues to be generated and is no longer consumed by the yeast, glucose remaining in the dough during baking contributes to crust browning and also to an increase in bread sweetness [18].

This enzyme has limited action on α-1,6 linkages, overriding side chains. However, some theories state that amyloglucosidase completely converts starch molecules to glucose [18].

2.2. Substrate: proteins

Proteins are composed of sequences of amino acids linked by peptide bonds. The main proteins of wheat flour are gliadin (a prolamine) and glutenin (a glutelin), which form, in the presence of water and mechanical energy, a cohesive protein network called gluten. This structure is very important for breadmaking. It has special viscoelastic properties (extensibility and elasticity) that allow the dough to flow. At the same time, it is able to retain CO_2 generated by the yeast during the fermentation step [18].

2.2.1. Glucose-oxidase (EC 1.1.3.4)

Glucose-oxidase converts glucose (from the hydrolysis of starch) and oxygen (present inside the dough) into gluconolactone and hydrogen peroxide (H_2O_2). The gluconolactone is natural and spontaneously converted to gluconic acid. H_2O_2 readily oxidizes the free thiol (—SH) groups of wheat flour dough proteins, promoting the formation of disulfide bonds (S—S) between gliadin and/or glutenin, that strengthen the gluten network. Thus, this enzyme is very important for breadmaking [18].

The cross-linking effect of proteins is responsible for the gluten network strengthening, that contributes for better crumb structure and bread volume improvement [18]. Nevertheless, high dosages of glucose-oxidase produce excessive stiffness of the dough reducing machinability, and must be avoided [19].

2.2.2. Hexose-oxidase (EC 1.1.3.5)

This kind of oxidoreductase has similar effects to those of glucose-oxidase. However, most widely, its substrates are mono and oligosaccharides, other than glucose. The corresponding lactones are obtained, and the generated H_2O_2 acts exactly the same way as in the case of glucose-oxidase as described under Section 2.2.1, producing similar effects in breadmaking [18].

2.2.3. Transglutaminase (EC 2.3.2.13)

This kind of acyl transferase promotes the reaction between amines, such as those presented by the γ-carboxamide from L-glutamine with the ε-amino group from L-lysine. This enzyme catalyzes the formation of covalent cross-linkages between proteins having these amino acid residues. It gives an additional strengthening effect to the gluten network comprising disulfide bonds. The result is the formation of larger and insoluble gluten polymers that affect not only the biochemical characteristics of the dough, but also its rheological properties [24]. Such an effect permits to replace the use of oxidants and even chemical emulsifiers in bakery formulations. Thus, transglutaminase is sometimes recommended in high-fiber and rye bread production. Gluten-free baked goods are also a promising field of action, as the utilization of transglutaminase enhances the protein network formation in breadmaking [18].

This enzyme increases water absorption of wheat flour doughs, provokes dough strengthening, enhances dough stability, reduces dough extensibility, improving crumb texture and bread volume [18].

Transglutaminase is recommended for reinforcing weak protein networks, and also for enhancing freeze-thaw stability of frozen doughs, like frozen croissants and puff pastry, as it decreases their deterioration during frozen storage [18].

2.2.4. Protease

Proteins present in baking doughs are substrates for proteases, which hydrolyze peptide bonds irreversibly, in order to reduce mixing time of bread doughs, or to reduce the strength of biscuit doughs, improving their machinability [19]. The disulfide cross-linkages of gluten are not affected by proteases and thus remain intact. The extension of protease effects depends on the amount of enzyme added and on the period of time that it is allowed to work before its inactivation by oven temperatures or pH changes. The main results of protease action are: (i) increase in protein water solubility; (ii) decrease in dough viscosity; (iii) decrease in the average molecular weight of protein fractions; and, consequently, (iv) decrease in gluten complex elasticity [18, 20].

Neutral or sulfhydryl proteases have been used more effectively due to their active pH range (from 5 to 8), that fits the pH of the majority of breads and biscuit doughs. Almost all the fungal proteases from *Aspergillus oryzae* are neutral type, while vegetable proteases, like papain and bromelain, are sulfhydryl type [20].

For long fermentation times, like in saltine cracker production, the dough can reach pH 4 or lower, and in this case, acidic protease is better used. Otherwise, in soda cracker production, the dough rises up to the alkaline region after soda addition, making serine protease (trypsin) more effective for gluten breakdown. This kind of protease is extracted mainly from bacterial sources like *Bacillus subtilis* [20].

High levels of protease cause such gluten network weakening that produces the coarse texture desired for English muffins, or favors cookie dough flow in the oven. However, care must be taken to avoid excessive proteolysis in bread doughs, because weak gluten networks generate undesirable coarse texture and low bread volume [20].

In the sponge process, it is usual to add small amounts of protease at the beginning of mixing, allowing its action on the gluten network during the sponge fermentation. When fresh flour is incorporated to the sponge, the newly added flour is poorly hydrolyzed during dough mixing. This blend of hydrolyzed and almost non-hydrolyzed gluten generates good smooth dough in the mixer that permits a decrease in mixing time [20].

It is useful to add small amounts of protease in the straight dough process for pan bread, to avoid tight doughs that give incomplete pan filling, or to avoid undesirable breaking along the loaf side. Similarly, in the production of hamburger and hot-dog breads, the dough must flow to fill in the molds during the short fermentation time. The addition of small amounts of protease in the mixer improves dough flow and enhances bread shape and symmetry [20].

In pizza dough production, the make-up work to spread and round the dough into a thin layer becomes easier as a result of adding small amounts of protease during mixing. In this case, the enzyme is able to work during proofing time, adequately reducing the strength of the gluten network, avoiding dough contraction during sheeting and preserving the desired oven spring [20].

The amino acids released by the proteolytic action react with the reducing sugars at high temperatures in the so-called Maillard reaction, enhancing color and flavor of breads and biscuits [18].

2.3. Substrate: lipids

Wheat flour lipids are composed of high levels of linoleic acid (C18:2), and lower levels of palmitic (C16:0) and oleic (C18:1) acids. These fatty acids may occur in the free form, or bound to starch and proteins. Starch lipids, mainly lysophospholipids, form complexes with amylose during gelatinization and have little importance for breadmaking [18].

Non-starch lipids (NSLs) (75% of total wheat flour lipids) are divided 1:1 into polar and non-polar lipids. Most of bound NSLs are composed by triacylglycerols (non-polar). Free NSLs are mainly composed of glycolipids and phospholipids; both are polar molecules that positively contribute to dough handling properties. They have a great effect on loaf volume, due to their effect on the stability of the gas cells, as they can form thin lipid monolayers inside gas cells that enhance CO_2 retention by the dough [18].

2.3.1. *Phospholipase (EC 3.1.4.3)*

Phospholipases are a particular kind of lipase with higher specificity toward phospholipids (polar fraction), that converts them *in situ* into lipids with even higher polarity and surface activity [25]. These act as dough strengthening emulsifiers, with dough stabilizing properties [18]. With the use of phospholipases, traditional emulsifiers like DATEM, CSL and SSL can be completely or partially substituted in breadmaking with similar results [25]. Phospholipases also improve dough machinability, as the stickiness is reduced, and the bread volume ultimately increases [18].

2.3.2. *Glycolipase (E.C. 3.1.1.26)*

Glycolipases are a particular kind of lipase with higher specificity toward glycolipids (polar fraction), that, similarly to phospholipase, converts them *in situ* into emulsifiers. Having similar effects in breadmaking as those from phospholipases, these enzymes increase dough stability. This effect is possible once the generated surface-active lipids maintain stable gas cell structures, due to the interaction of polar lipids with proteins at the liquid lamellae that surround gas cells [25].

2.3.3. *Lipase (EC 3.1.1.3)*

This kind of enzyme is classified as a glycerol ester hydrolase due to its capacity to hydrolyze acylglycerol ester linkages, releasing preferably fatty acids at positions −1 and −3 from the glycerol structure. The products formed include mono- and diacylglycerol residues, which act as crumb softening emulsifiers in breadmaking. This effect is due to the acylglycerols capacity to penetrate amylose helicoidal structure forming amylose-lipid complexes, retarding amylose retrogradation, increasing bread volume and providing better crumb structure and texture [18].

2.3.4. *Lipoxygenase (EC 1.13.11.12)*

The substrates of lipoxygenase are polyunsaturated fatty acids, such as linoleic (C18:2) and linolenic (C18:3) acids, and β-carotene and chlorophylls from wheat flour [18, 19].

This enzyme, present in enzyme-active soy flour, oxidizes endogenous wheat flour pigments, providing a bleaching effect, resulting in a whiter crumb. Also, dough strengthening occurs during breadmaking [20]. The accessible thiol (—SH) groups from wheat flour proteins are oxidized by the hydroxyperoxides generated by lipoxygenase action on fatty acids. This oxidation provokes intermolecular disulfide bond formation among gluten proteins, increasing mixing tolerance, improving dough machinability, enhancing rheological properties for breadmaking, increasing bread volume and improving internal texture. Nevertheless, high dosages of lipoxygenase produce undesirable flavors in breads, due to the decomposition of the hydroxyperoxides of fatty acids generated by lipoxygenase action, and must be avoided [18].

2.4. Substrate: non-starch polysaccharides (NSPS)

There are several non-starch polysaccharides (NSPS) in wheat flour: pentosans, β-glucans and cellulose, all classified as dietary fiber constituents [18]. Pentosans are the most important NSPS due to their great water absorption capacity, despite their low content (2–3%) in wheat flour.

Around 50% of pentosans are water soluble, and 50% insoluble. About 75% of pentosans are xylans, and almost 25% are galactans. Due to their strong hydrophilicity, pentosans affect dough viscosity and, consequently, loaf volume [20].

Xylans are xylose polymers linked by β-1,4 bonds. They can have arabinose molecules linked to the xylan backbone by β-1,3 bonds; then, they are called arabinoxylans (AXs). Some linkages can be β-1,2, mainly in the insoluble or water unextractable arabinoxylans (WU-AXs). Soluble or water extractable arabinoxylans (WE-AXs) present a 3:1 xylose:arabinose ratio, while WU-AXs have a greater proportion of arabinose [20].

AXs are the main NSPS that constitute wheat endosperm cell walls, and, in solution, provide high viscosities, which depend on AXs molecule length. Both WE-AXs and WU-AXs have great water-binding capacity, which, in breadmaking, increases dough consistency, stiffness and resistance to extension, while decreasing mixing time and dough extensibility [18].

The WE-AXs are weakly linked to wheat endosperm cell walls and have gelling properties in the presence of oxidants [25]. The main components responsible for the increase in viscosity of flour suspensions are the WE-AXs, and this ability stabilizes protein films during temperature elevation [18]. According to Wang et al. [27], WE-AXs are considered beneficial to bread quality, enhancing gas retention.

The WU-AXs are structural components of wheat cell walls that link AXs, proteins, cellulose and lignin, through covalent and non-covalent bonds [26]. Experiments have shown better loaf volume and bread quality when WU-AX content decreases, and this effect is due to: (i) physical barriers to gluten development represented by the WU-AX, which impair gliadin and glutenin approximation; (ii) high water absorption capacity, making water unavailable for gluten network development; and (iii) gas cell perforation by these structures, provoking their coalescence [18, 27].

If the AXs do not receive appropriate enzymatic treatment during dough processing, the water added to the wheat flour becomes constrained in these hydrophilic structures, causing a water scarcity for gluten network development, enzyme action, yeast activity and starch granule gelatinization, impairing bread final quality.

2.4.1. Fungal xylanase (EC 3.2.1.8)

This enzyme is used to release water from xylans. It has great influence on dough viscosity. Thus, it improves dough tolerance to the breadmaking processes, as dough elasticity is reduced [19]; and contributes to increase bread volume up to 20% when compared with a control, mainly in high-fiber doughs, such as breads made with whole wheat flour and other whole cereals [20]. Xylanases enhance gas retention capacity of dough, contributing to a softer and finer crumb [19]. This kind of endo-xylanase is extracted from *Aspergillus* spp. and this enzyme preferentially hydrolyzes WE-AX, promoting gluten protein aggregation [26], due to its water releasing capacity which is beneficial for gluten network formation [19].

Excessive dosage levels must be avoided, because, in this case, slack and sticky wheat flour doughs are produced. This effect is caused by the excessive hydrolysis of AX, provoking excessive loss in water binding capacity [19]. The resultant breads present in appropriate crumb structure, with ragged gas cell distribution, besides inappropriate crust color [18].

2.4.2. Bacterial xylanase (EC 3.2.1.55)

This kind of endo-xylanase is extracted from *B. subtilis*. It preferentially hydrolyzes WU-AX, enhancing dough stability. Due to this effect, the dough is able to keep maximum volume for a longer period during the fermentation step, and it maintains a great resistance to mechanical stress during the breadmaking process [18]. Oven spring is prolonged and bread volume is enhanced due to dough relaxation and better gas retention [19, 26], which produces finer grains that provide a softer and more homogeneous bread crumb [18].

For the same reason as for fungal xylanase, excessive dosage levels of bacterial xylanase must also be avoided [18].

2.4.3. Cellulase (EC 3.2.1.4)

This enzyme hydrolyzes cellulose (linear homopolysaccharide composed by a glucose polymer backbone linked by β-1,4 bonds) from wheat cell walls, mainly from the wheat grain outer layers. Cellulose chains are organized in crystalline and amorphous regions. In cellulose crystalline structure, the molecules are highly ordered and chain arrangement blocks water and enzyme penetration into the microfibrils. In the non-crystalline (amorphous) regions, water and enzymes have greater access, and these sites are more easily hydrolyzed than the crystalline ones. Thus, the amorphous regions are firstly attacked and degraded by the cellulases [28]. This produces lower molecular weight fragments that can bind more water.

Cellulase action on cellulose has numerous benefits in the breadmaking process: (i) water absorption increases; (ii) dough viscosity increases; (iii) high-fiber dough stickiness decreases; (iv) machinability is enhanced; (v) the release of glucose increases, and; (vi) the cut opening for French rolls increases [28].

3. Future trends

There is currently huge pressure on the food industry to produce healthier products. "Clean" or "friendly" labels, with shorter and simpler ingredient lists are a strong trend. These include the search for more natural and healthier alternatives for chemical additives which have a negative impact on consumer acceptance. The bakery industry is trying to eliminate E-number ingredients from its formulations using, for example, (i) enzymes and vital wheat gluten (an ingredient) to eliminate emulsifiers and chemical oxidants; (ii) hydrocolloids as a more "friendly" choice than other additives; and (iii) natural preservatives such as fermentates, for mold control. However, in some cases, these alternatives are expensive and not as effective as chemical additives.

Enzymes do not need to be declared as processing aids on the labels of food products in many countries, so they are an interesting strategy for "clean labels". Some enzymes are under development and will probably soon become commercially available for use in breadmaking. An example is laccase (EC 1.10.3.2), an oxidative enzyme that oxidizes different kinds of phenolic compounds, increasing dough stability and strength, promoting quicker dough formation and reducing dough stickiness [18]. Another example is β-glucanase (EC 3.2.1.6), which hydrolyzes the β-glucans present in barley, rye and oat flours, enhancing microstructure, volume, texture, shelf-life and taste in breads made with these composite flours [29].

Author details

Luis Carlos Gioia, José Ricardo Ganancio and Caroline Joy Steel*

*Address all correspondence to: steel@unicamp.br

Department of Food Technology, School of Food Engineering, University of Campinas (UNICAMP), Campinas, SP, Brazil

References

[1] Msagati TAM. The Chemistry of Food Additives and Preservatives. 1st ed. Chichester: Wiley-Blackwell; 2013. 322 p

[2] De Leyn I. Other Functional Additives. In: Zhou W, editor. Bakery Products Science and Technology. 2nd ed. Chichester: Wiley-Blackwell; 2014. pp. 295-306

[3] Zhou W, Therdthai N, Hui YH. Introduction to baking and bakery products. In: Zhou W, editor. Bakery Products Science and Technology. 2nd ed. Chichester: Wiley-Blackwell; 2014. pp. 3-16

[4] Sahi SS. Ascorbic acid and redox agents in bakery systems. In: Zhou W, editor. Bakery Products Science and Technology. 2nd ed. Chichester: Wiley-Blackwell; 2014. pp. 183-197

[5] Cauvain S. Technology of Breadmaking. 3rd ed. Switzerland: Springer; 2015. 408 p. DOI: 10.1007/9783319146874

[6] Moonen H, Bas H. Mono- and diglycerides. In: Norn V, editor. Emulsifiers in Food Technology. 2nd ed. Chichester: Wiley-Blackwell; 2015. pp. 73-91

[7] Orthoefer F. Applications of emulsifiers in baked foods. In: Hasenhuettl GL, Hartel RW, editors. Food Emulsifier and Their Applications. 2nd ed. New York: Springer; 2008. pp. 263-284. DOI: 10.1007/9780387752846.ch9

[8] Gaupp R, Adams W. Diacetyl Tartaric Acids of Monoglycerides (DATEM) and associated emulsifiers in bread making. In: Norn V, editor. Emulsifiers in Food Technology. 2nd ed. Chichester: Wiley-Blackwell; 2015. pp. 121-145

[9] Boutte T, Skogerson L. Stearoyl-2-lactylates and oleoyl lactylates. In: Norn V, editor. Emulsifiers in Food Technology. 2nd ed. Chichester: Wiley-Blackwell; 2015. pp. 251-270

[10] Cottrell T, van Peij J. Sorbitan esters and polysorbates. In: Norn V, editor. Emulsifiers in Food Technology. 2nd ed. Chichester: Wiley-Blackwell; 2015. pp. 271-295

[11] Ferrero C. Hydrocolloids in wheat breadmaking: A concise review. Food Hydrocolloids. 2017;**68**:15-22. DOI: 10.1016/j.foodhyd.2016.11.044

[12] Habibi H, Khosravi-Darani K. Effective variables on production and structure of xanthan gum and its food applications: A review. Biocatalysis and Agricultural Biotechnology. 2017;**10**:130-140. DOI: 10.1016/j.bcab.2017.02.013

[13] Thombarea N, Jha U, Mishra S, Siddiqui MZ. Guar gum as a promising starting material for diverse applications: A review. International Journal of Biological Macromolecules. 2016;**88**:361-372. DOI: 10.1016/j.ijbiomac.2016.04.001 0141-8130

[14] Suhr KI, Nielsen PV. Effect of weak acid preservatives on growth of bakery product spoilage fungi at different water activities and pH values. International Journal of Food Microbiology. 2004;**95**:67-78. DOI: 10.1016/j.ijfoodmicro.2004.02.004

[15] Legan JD. Mould spoilage of bread: The problem and some solutions. International Biodeterioration & Biodegradation. 1993;**32**:33-53. DOI: 10.1016/0964-8305(93)90038-4

[16] Samapundo S, Devlieghere F, Vroman A, Eeckhout M. Antifungal activity of fermentates and their potential to replace propionate in bread. LWT-Food Science and Technology. 2017;**76**:101-107. DOI: 10.1016/j.lwt.2016.10.043

[17] Enzyme nomenclature-Recommendations of the Nomenclature Committee of the International Union of Biochemistry and Molecular Biology (NC-IUBMB) on the nomenclature and classification of enzymes by the reactions they catalyze [Internet]. 2017. Available from: http://www.chem.qmul.ac.uk/iubmb/enzyme/ [Accessed: 06-03-2017]

[18] Goesaert H, Gebruers K, Courtin CM, Brijs K, Delcour JA. Enzymes in breadmaking. In: Hui YH, Corke H, Leyn ID, Nip WK, Cross N, editors. Bakery Products Science and Technology. Ames: Blackwell Publishing; 2006. pp. 337-364

[19] Sluimer P. Principles of Breadmaking. Minnesota, USA: American Association of cereal Chemists, Inc.; 2005. 212 p

[20] Stauffer CE. Functional Additives for Bakery Foods. New York, USA: AVI Book; 1990. 280 p

[21] Whitehurst RJ, Law BA. Enzymes in Food Technology. Boca Raton: Sheffield Academic Press. 2002. 255 p

[22] Gomes-Ruffi CR, Cunha RH, Almeida EL, Chang YK, Steel CJ. Effect of the emulsifier sodium stearoyl lactylate and of the enzyme maltogenic amylase on the quality of pan

bread during storage. LWT-Food Science and Technology, 2012;**49**:96-101. DOI: 10.1016/j.lwt.2012.04.014

[23] Gerrard JA, Every D, Sutton KH, Gilpin MJ. The role of maltodextrins in the staling of bread. Journal of Cereal Science. 1997;**26**:201-209. DOI: 10.1006/jcrs.1997.0121

[24] Autio K, Kruus K, Knaapila A, Gerber N, Flander L, Buchert J. Kinetics of transglutaminase-induced cross-linking of wheat proteins in dough. Journal of Agricultural and Food Chemistry. 2005;**53**:1039-1045. DOI: 10.1021/jf0485032

[25] Almeida EL, Chang YK. Effect of the addition of enzymes on the quality of frozen pre-baked French bread substituted with whole wheat flour. LWT-Food Science and Technology. 2012;**49**:64-72. DOI: 10.1016/j.lwt.2012.04.019

[26] Courtin CM, Delcour JA. Relative activity of endoxylanases towards water-extractable and water-unextractable arabinoxylan. Journal of Cereal Science. 2001;**33**:301-312. DOI: 10.1006/jcrs.2000.0354

[27] Wang M, Hamer RJ, Van Vliet T, Gruppen H, Marseille H, Weegels PL. Effect of water unextractable solids on gluten formation and properties: mechanistic considerations. Journal of Cereal Science. 2003;**37**:55-64. DOI: 10.1006/jcrs.2002.0478

[28] Santos FR da S. Production and characterization of cellulases and hemicellulases by mesofilic fungal strains isolated from South Mato-Grosso state's Cerrado [thesis]. Dourados, Brazil: Federal University of Grande Dourados; 2014

[29] Li Z, Dong Y, Xinghua Z, Xiao X, Zhao Y, Yu L. Dough properties and bread quality of wheat-barley composite flour as affected by β-glunacase. Cereal Chemistry. Nov./Dec. 2014;**91**(6):631. DOI: 10.1094/CCHEM-01-14-0019-R